Enhanced Oil Recovery: Field Planning and Development Strategies

Enhanced Oil Recovery: Field Planning and Development Strategies

Editor

Suhas Kulkarni

scitus
academics

Enhanced Oil Recovery: Field Planning and Development Strategies

Edited by **Suhas Kulkarni**

Printed in 2017

ISBN: 978-1-68117-364-1

Library of Congress Control Number: 2015936528

© 2016 by
SCITUS Academics LLC,
616, Corporate Way, Suite 2, 4766,
Valley Cottage, NY 10989

www.scitusacademics.com

Contents

Preface

Enhanced-Oil Recovery (EOR) evaluations focused on asset acquisition or rejuvenation involve a combination of complex decisions, using different data sources. EOR projects have been traditionally associated with high CAPEX and OPEX, as well as high financial risk, which tend to limit the number of EOR projects launched. In this book, the authors propose workflows for EOR evaluations that account for different volumes and quality of information. This flexible workflow has been successfully applied to oil property evaluations and EOR feasibility studies in many oil reservoirs. The methodology associated with the workflow relies on traditional (look-up tables, XY correlations, etc.) and more advanced (data mining for analog reservoir search and geology indicators) screening methods, emphasizing identification of analogues to support decision making. The screening phase is combined with analytical or simplified numerical simulations to estimate full-field performance by using reservoir data-driven segmentation procedures.

Editor

Advances in Asset Management Techniques: An Overview of Corrosion Mechanisms and Mitigation Strategies for Oil and Gas Pipelines

Chinedu I. Ossai

Production Planning Department, Overall Forge Pty Ltd, 70 R W Henry Drive, Ettamogah near Albury, Albury, NSW 2640, Australia

ABSTRACT

Effective management of assets in the oil and gas industry is vital in ensuring equipment availability, increased output, reduced maintenance cost, and minimal nonproductive time (NPT). Due to the high cost of assets used in oil and gas production, there is a need to enhance performance through good assets management techniques. This involves the minimization of NPT which accounts for about 20–30% of operation time needed from exploration to production.

Corrosion contributes to about 25% of failures experienced in oil and gas production industry, while more than 50% of this failure is associated with sweet and sour corrosions in pipelines. This major risk in oil and gas production requires the understanding of the failure mechanism and procedures for assessment and control. For reduced pipeline failure and enhanced life cycle, corrosion experts should understand the mechanisms of corrosion, the risk assessment criteria, and mitigation strategies. This paper explores existing research in pipeline corrosion, in order to show the mechanisms, the risk assessment methodologies, and the framework for mitigation. The paper shows that corrosion in pipelines is combated at all stages of oil and gas production by incorporating field data information from previous fields into the new field's development process.

INTRODUCTION

The oil and gas industry is an asset intensive business with capital assets ranging from drilling rigs, offshore platforms and wells in the upstream segment, to pipeline, liquefied natural gas (LNG) terminals, and refineries in the midstream and downstream segments. These assets are complex and require enormous capital to acquire. An analysis of the five major oil and gas companies (BP, Shell, ConocoPhillips, Exxonmobil, and Total) shows that plant, property, and equipment on average accounts for 51% of the total assets with a value of over $100 billion [1]. Considering the huge investment in assets, oil and gas companies are always under immense pressure to properly manage them. To achieve this involves the use of different optimization strategies that is aimed at cost reduction and improved assets reliability [2].

Due to the growth in the demand of oil and gas around the world, companies are developing new techniques to reach new reservoirs in the offshore and onshore arena [3]. This is putting pressure on most of the facilities with the attendant cost of maintenance soaring [1]. The continuous utilization and the ageing of facilities have resulted in record failures in the oil and gas plants. Research shows that between 1980 and 2006, 50% of European, major hazards of loss containment events arising from technical plants failures were primarily due to ageing plants mechanism caused by corrosion, erosion, and fatigue [4, 5].

A study shows that corrosion cost in US rose above 1$ trillion in 2012 accounting for about 6.2% of GDP hence, the largest single expense in the economy [6]. In the oil and gas company, corrosion accounts for over 25% of assets failure [7] and is found to be prevalent in every stage of the production cycle. Oxygen which plays a dominant role in corrosion is normally present in producing formation water. During drilling operation, drilling mud can corrode the well casing, drilling equipment, pipeline, and the environment. Water and CO_2 produced or injected for secondary recovery can cause severe corrosion of completion strings, while the acids used to reduce formation damage around the well or to remove scale can attack metals [8]. The formation water and injected water used for the oil recovery are a potential source of pipeline corrosion during transportation of the oil from the wells to the loading terminals. Mechanical static equipment like valves, tanks, vessels, separators, and so forth are susceptible to a different kind of corrosion however, pipelines are more prone to corrosion due to the presence of CO_2, H_2S, H_2O, bacteria, sand, and so forth in the fluid.

Owing to the increasing cost of pipeline corrosion management in the oil and gas industries [1], operators are becoming more concerned about corrosion management planning at all phases of production. Corrosion information from existing field data is being incorporated into design information for new oil and gas field [9, 10] in a bid to develop appropriate corrosion management methodologies that will enhance the design life of the pipelines and optimize production. To reduce the risk of microbiologically influenced Corrosion (MIC) and other associated corrosions like stress corrosion cracking (SCC), hydrostatic testing of carbon steel pipes should be carried out in such a manner that enhances the future pipeline service conditions by using the right source of water, ensuring proper degree of filtration, ensuring limited exposure period to temperature and eliminating air packets [11]. Though bacteria in the biofilm are responsible for pitting of a pipeline in a MIC however, the impact of the flow velocity of the constituent fluid influences the mass transfer rate thereby affecting the biofilm formation, hence, inhibiting the activities of sulphate reducing bacteria, (SRB) present in the fluid [12]. This flow attribute has significant impact in MIC in oil and gas pipeline.

Considering the fact that the CO_2 and H_2S induced corrosion rate can reach up to 6 mm/yr and 300 mm/yr, respectively, [13] in oil and gas pipelines, sophistication in inspection and monitoring techniques is therefore necessary for quick mitigation. The increased trend in in-line inspection and online data acquisition has helped in quicker data acquisition, analysis, and decision making regarding corrosion in pipelines. The enhanced research knowledge of the behaviour of these corrodents (CO_2 and H_2S, acetic acid, etc.) at different operating conditions [14–17] has given rise to numerous mechanistic, statistical, and empirical models [18–23] which have contributed immensely in the inspection and monitoring, selection of inhibitors, and materials selection for pipelines design.

Since corrosion is a dominant factor contributing to failures and leaks in pipelines [24], to aid industry experts in managing the integrity of pipelines therefore involves a layout of the developments in the management strategies. This involves the recognition of the conditions contributing to the corrosion incident and identifying effective measures that can be taken to mitigate against them. To facilitate best practices in pipeline integrity management therefore, requires a framework that utilizes good policies and procedures in inspection, data collection, and interpretation for corrosion control.

OVERVIEW OF CORROSION

Corrosion is a naturally occurring phenomena commonly defined as the deterioration of a substance (usually metal) or its properties because of a reaction with its environment [25]. Corrosion of materials is inevitable due to the fundamental need of lowering of Gibbs energy [26]. Every material is trying to achieve a lower energy state hence the ability to corrode in order to get to a low energy oxide state. Though this is the case with all materials, the major focus of experts however, is to achieve an equilibrium position between the materials and the environment thereby controlling corrosion.

Modern corrosion science has its roots in electrochemistry and metallurgy. Whereas electrochemistry contributes to the understanding of materials via corrosion, metallurgy provides information about the behaviour of the material and their alloys hence provide a medium for combating the degradation on them. The type of corrosion mechanism and its rate of attack depend on the nature of the environment (air, soil, water, etc.) in which the corrosion takes place. Whereas some environmental condition can help to mitigate the rate of corrosion, others help to increase it hence, industrial wastes and products can either be corrosion inhibitor or catalyst. For instance, CO_2, H_2S, temperature, mass flow rate, pH, formation water, and so forth contribute in no small measure to the rate of corrosion in oil and gas pipeline [14, 16, 17, 27]. The existence of anodic cathodic sites on the surface of a piece of metal implies that the difference in electrical potential is found on the surface. This potential difference has the tendency of initiating corrosion. If an oil and gas pipeline passes through a zone of clay soil (where the oxygen concentration is low) to gravel (where the oxygen concentration is high), the part of the pipeline in contact with the clay becomes anodic and suffers damage. Though this problem is extensively addressed with the cathodic protection [26], concentration cell may also be formed where there are differences in metal ion concentration.

Although most metals are crystalline in form, they generally are not continuous single crystal but rather are collections of small grains of domains of localized order in which microcrystal forms as the liquid cools and solidifies. In the final states, the crystals have different orientation with respect to one another. The edge of the domain form grain boundaries which are an example of planar defects in metal. These defects are usually sites of chemical reactivity. The boundaries are also weaknesses, the places where stress corrosion cracking begins. The metallic surface exposed to an aqueous electrolyte usually possesses site for oxidation (anodic reaction) that produces electrons in the metal and reduction (cathodic reaction) that consumes the electrons produced by the anodic reaction [25, 26]. These sites make up a corrosion cell. The anodic reaction (Figure 1) involves the dissociation of metal to form either soluble ionic product or an insoluble compound of metal usually an oxide. For cathodic reaction (Figure 2), oxygen gas generated could be reduced or water is reduced to produce hydrogen gas. The simultaneous reaction of the anodic and cathodic reactions produces the electrochemical cell.

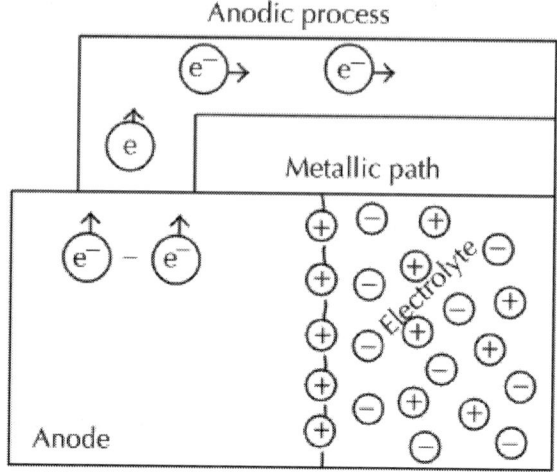

Figure 1: Anodic process.

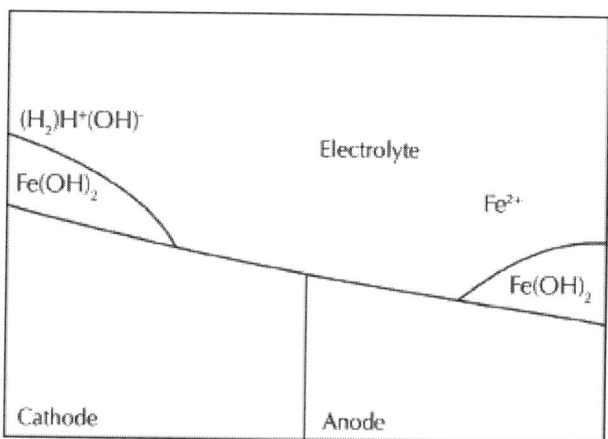

Figure 2: Cathodic process.

In completely oxygen-free water, the cathodic reaction that takes place is the reaction of hydrogen ion to form hydrogen gas as shown in (1):

$$2H^+ + 2e^- \rightarrow H_2(g)$$

(1)

When significant amounts of oxygen are present in the system, the cathodic reaction that takes place is shown in (2):

$$2H^+ + \frac{1}{2}O_2 + 2e^- \rightarrow H_2O$$

(2)

The hydrogen ion is present in water due to the ubiquitous dissolution of water into hydroxyl ions as shown in (3):

$$2H_2O \rightarrow 2H^+ + 2(OH)$$

(3)

In the anode, there is a dissociation of iron to form a ferrous ion as shown in (4).

$$Fe \rightarrow Fe^{2+} + 2e$$

(4)

The ferrous ion will react with the hydroxyl ion to form insoluble ferrous hydroxide as shown in (5):

$$Fe^{2+} + 2(OH)^- \rightarrow Fe(OH)_2$$

(5)

The anodic and cathodic reactions that take place in a neutral and alkaline condition are shown in Figure 3.

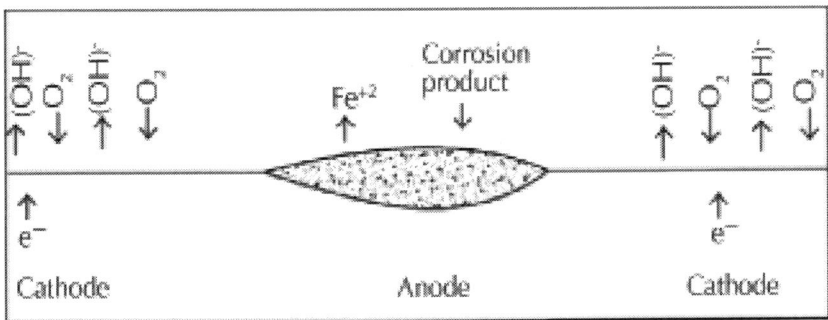

Figure 3: Neutral and Alkaline condition of a corrosion process.

The cathodic reaction is as follows:

$$\frac{1}{2}O_2 + H_2O + 2e^- \rightarrow 2(OH)$$

(6)

while the anodic reaction is the same as (4).

For an anodic condition, the cathodic and anodic reactions are represented in Figure 4.

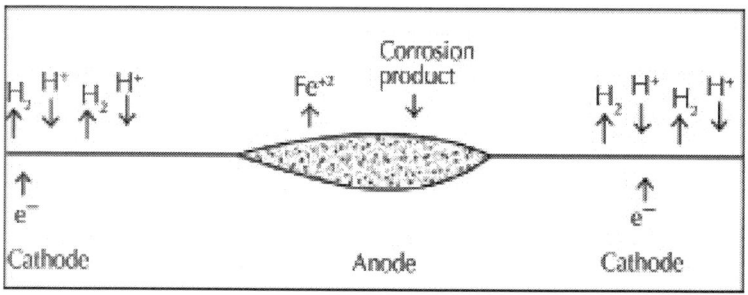

Figure 4: Acidic Condition of a Corrosion process.

The cathodic reaction equation is shown in (1), while the anodic equation is shown in (4).

In a deoxygenated solution, the hydrogen reaction combines with the others to yield the net corrosion reaction shown in (7):

$$Fe + 2H_2O \rightarrow Fe(OH)_2 + H_2(g)$$

(7)

In oxygenated aqueous systems, the oxygen reduction leads to a slightly different net corrosion reaction as shown in (8):

$$Fe + H_2O + \frac{1}{2}O_2 \rightarrow Fe(OH)_2$$

(8)

Whereas in deoxygenated solution, hydrogen is evolved, in oxygenated system, oxygen is consumed. The evolved hydrogen acts as a catalyst for the formation of magnetite (Fe_3O_4) in deoxygenated water. Experiment shows that hydroxide readily decomposes into magnetite in deoxygenated water above 100°C [28] as indicated in (9):

$$3Fe(OH)_2 \rightarrow Fe_3O_4 + H_2\,(g) + 2H_2O$$

(9)

The net corrosion reaction with the magnetite as the final product is shown in (10):

$$3Fe + 4H_2O \rightarrow Fe_3O_4 + 4H_2\,(g)$$

(10)

In oxygenated solution, the ferrous oxide (Fe^{2+}) does not immediately precipitate out since it rapidly oxidizes to ferric oxide (Fe^{3+}), as a result, insoluble iron hydroxide is formed which is converted to hematite as shown in (11):

$$Fe\,(OH)_2 + \frac{1}{2}H_2O + \frac{1}{4}O_2 \rightarrow Fe\,(OH)_3$$

(11)

The ferric oxide (Fe^{3+}) is converted to magnetite according to (12)

$$2Fe\,(OH)_3 + Fe\,(OH)_2 \rightarrow Fe_3O_4 + 4H_2O$$

(12)

Cathodic and anodic sites could be built as a result of variation in environmental conditions, metallic microstructure variation, and variation in environmental concentration of oxygen at different points of a metal [26]. At the anodic sites, the dissolution of metallic ions in the electrolyte brings about the flow of electrons between the corroding anodes and non-corroding cathodes. The spontaneous nature of the corrosion however, depends on the rate of flow of these electrons.

Though establishing the tendency for corrosion is necessary, however, it is more important to determine the rate of corrosion. This is because a particular metal or alloy may be prone to corrosion in an environment but at a very low rate, in which it will not be a problem [26]. To understand the rate of corrosion however, requires the knowledge of the role of primary environment and metallurgical variables, underlying mechanism of corrosion, and synthesis of information to account for effects of the parameters.

MECHANISMS OF CORROSION IN OIL AND GAS PIPELINES

Fluid flowing from oil and gas pipelines has a combination of chemicals including CO_2, H_2S, organic acids, bacteria, sand, and water. These constituents are among the major causes of corrosion in pipeline. The CO_2 dissolves in the presence of water to form an acidic oxide which reacts with iron This type of corrosion is referred to as sweet corrosion. This is responsible for most types of general corrosion in oil and gas pipeline. Sour corrosion occurs when H_2S in the excess of 100ppm is present in the oil and gas, causes corrosion in the pipeline, and predominantly causes pitting [26, 29].

CO_2 present in oil and gas will dissolve in water to produce carbonic acid (H_2CO_3) [23, 27]. This acid dissolves steel to produce iron carbonate and hydrogen as shown in (13). This reaction takes place at the cathode:

$$Fe + H_2CO_3 \rightarrow FeCO_3 + H_2\,(g)$$

(13)

Despite the weakness of carbonic acid it is extremely corrosive to carbon steel. The chemical reactions above form the iron carbonate films. Depending on the condition during the formation, these films can be protective or non-protective at the anode, iron dissolves as shown in (4). The presence of CO_2 acts as a catalyst increasing the hydrogen evolution thereby increasing the corrosion rate of carbon steel in aqueous solution [27] The carbonic acid (H_2CO_3) either serves as an extra source of H^+ or is reduced directly according to (14) and (15):

$$2H^+ + 2e^- \rightarrow H_2\,(g)$$

(14)

$$2H_2CO_3 + 2e^- \rightarrow H_2\,(g) + 2HCO_3^-$$

(15)

The dissolved iron concentration will increase until Fe^{2+} is the same as the precipitation rate of $FeCO_3$ [30]. When Fe^{2+} is released in the corrosion process, the double amount of bicarbonate forms according to (16):

$$Fe + 2H_2CO_3 \rightarrow Fe^{2+} + H_2 + 2HCO_3$$

$$(16)$$

The pH increases until bicarbonate and carbonate becomes so high that solid $FeCO_3$ precipitates [30] as shown in equation (17):

$$Fe^{2+} + 2HCO_3^- \rightarrow FeCO_3 (s) + H_2CO_3$$

$$(17)$$

When all the ferrous ions produced by corrosion precipitates as iron carbonate ($FeCO_3$), the pH remains constant and the overall reaction becomes as the state in (13).

In order to control the rate of corrosion on the pipeline, there should be passivity. Passivity is the condition existing on a metal surface because of the presence of protective film. When protective film is formed on the metal surface, it forms a coat which hinders further corrosion action on the material [26, 31]. The structure of the passive film (magnetite) formed on low carbon steel oxidizes in high temperature and has two distinct layers on the steel. The inner layer is compact and adheres well to the steel and has uniform thickness. The outer layer is a porous mass of individual crystal that would flake off the steel in some place and very nonuniform in thickness (Figure 5). This protective film is removed from the surface of the pipeline through erosion, dissolution, and turbulence resulting in more corrosion. The possible mechanisms resulting in the removal of the protective film are a follows:

1. Dissolution or removal of protective layer by hydrodynamic shear stress occurs when the shear stress is greater than the bonding force between the film and the substrate. This is a function of a mechanical process of erosion caused by the multiphase flow regime in pipeline [19, 32].

2. In distributed flow condition, local near wall density of turbulence helps to remove the protective film. This disruption to the mass transfer boundary layer results in an enhanced corrosion rate [32, 33].

3,. Dissolution of film which is controlled by mass transfer. Thus the breakaway velocity may reflect conditions where the dissolution rate of the film is greater than the growth rate of the film [34].

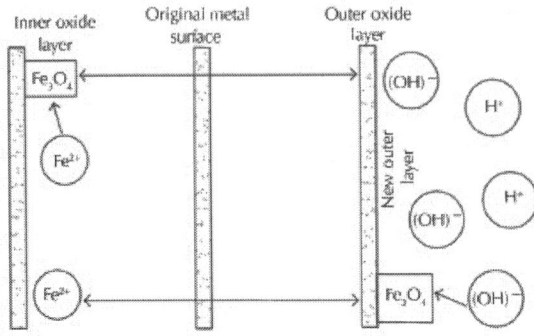

Figure 5: Schematic of magnetite double layer showing oxide formation locations.

The breakdown of protective film leads to the formation of localized corrosion that results in some of the major sources of corrosion failures like pitting, crevice, intergranular, and stress corrosion [12, 26, 35, 36]. The predominant breakdown processes are electrochemical and mechanical. Mechanical breakdown occurs when the protective film is ruptured as a result of stress or abrasive wear while, the electrochemical breakdown is a function of chemical reaction between the fluid constituent and the steel.

Types of Corrosion in Oil and Gas Pipeline

The primary chemical components that cause corrosion reaction to occur in pipeline are oxygen, acidic sulphur, and acidic chloride that dissolves in the water in the pipeline. The mechanism present in a given piping system varies according to the fluid composition, service location, geometry, temperature and so forth. In all cases of corrosion, the electrolyte must be present for the reaction to occur.

Internal Corrosion

Internal corrosion has become an increasing problem in most oil and gas pipelines as water cuts have increased and previously oil wet pipe surfaces have become water wet (providing the electrolyte for the corrosion cell) and as bacterial activities increases in the production system. Internal corrosion is the largest cause of pipeline failure in oil and gas industries [24] through different forms of corrosion like microbiologically influenced corrosion (MIC), erosion (flow enhanced) corrosion, under deposit (concentration cell) corrosion and so forth.

Erosion-Corrosion

The erosion-corrosion mechanism increases corrosion reaction rate by continuously removing the passive layer of corrosion products from the wall of the pipe. The passive layer is a thin film of corrosion product that actually serves to stabilize the corrosion reaction and slow it down. As a result of the turbulence and high shear stress in the line, this passive layer can be removed causing the corrosion rate to increase [37]. The erosion-corrosion is always experienced where there is high turbulence flow regime with significantly higher rate of corrosion than just corrosion or erosion in pipeline [38]. In a multiphase flow regime with a fully developed turbulent flow, bubbles development and collapse have been attributed to changes in mass transfer coefficient and an eventual increase in CO_2 corrosion in pipeline [34].

Under Deposit Corrosion

The under deposit mechanism can increase the corrosion reaction rate by causing a localized chemical concentration which results in pitting of the metal surface under solid deposits. These deposits appear to be composed of a corrosion product matrix with entrapment of formation solids, sand, and iron sulphide. The rate of corrosion under this mechanism is significantly lower than erosion-corrosion mechanism.

Microbiologically Induced Corrosion (MIC)

This type of corrosion is caused by bacterial activities. The bacteria produce waste products like CO_2, H_2S and organic acids that corrode the pipes by increasing the toxicity of the flowing fluid in the pipeline. Some bacteria like sulphate removing bacteria (SRB) consume hydrogen that is a product in a standard corrosion reaction process. This activity causes the existing corrosion rate to increase in an attempt to reach reaction equilibrium by replacing the hydrogen consumed by bacteria. Bacteria also accumulate on the pipe walls, creating deposits and under deposit corrosion. MIC is recognized by the appearance of black slimy waste material or nodules on the pipe surface as well as pitting of the pipe wall underneath these deposits.

Pitting Corrosion

Pitting is classified as a localized attack that results in rapid penetration and removal of metal at small discrete area. The initiation of a pit occurs when electrochemical or chemical breakdown exposes a small local site on a metal surface to damaging species such as chloride ion. The site where pitting occurs is where there is an environmental variation in comparison to the entire metal surface. The combination of chlorine with H_2S results in localized pitting on steel [35]. This area of pitting which is usually the anode normally get highly degraded due to enormous electron transfer between the entire large area of the metal surface which is the cathode and small anode (the pitting site).

Crevice Corrosion

Crevice corrosion results when a portion of a metal surface is shielded in such a way that the shielded portion has limited access to the surrounding environment. Such surrounding environment contain, damaging corrosion species usually chloride ion. A typical example of crevice corrosion is the crevice found at the area between two metal surfaces in close contact with a gasket or another metal surface. The environment that eventually forms in the crevice is similar to that formed under the precipitated corrosion that covers a pit. An electrochemical corrosion cell is formed from the couple between the unshielded surface and the crevice interior exposed to an environment with a lower oxygen concentration compared with the surrounding medium. The concentration of being the anode of a corrosion cell and existing in an acidic, high-chloride environment where repassivation is difficult makes the crevice interior subject of corrosion attack.

Stress Corrosion Cracking (SCC)

Stress corrosion cracking (SCC) is a form of localized corrosion which produces cracks in metals by simultaneous action of a corrodent and tensile stress. It propagates over a range of velocities from 10^{-3} -10 mm/h depending upon the combination of alloy and environment involved. The geometry is such that if they grow to appropriate lengths, they may reach a critical size that results in a transition from the relatively slow crack growth rate associated with stress corrosion to fast crack propagation rates associated with purely mechanical failure. This transition happens when the stress intensity, which is a function of the geometry of the component including the crack size, reaches the fracture value for the material concerned. SCC in pipeline is a type of environmentally associated cracking (EAC). This is because the crack is caused by various factors combined with the environment surrounding the pipe. The most obvious identifying characteristic of SCC in pipeline is high pH of the surrounding environment, appearance of patches, or colonies of parallel cracks on the external of the pipe [39].

Top of the Line Corrosion (TLC)

This type of corrosion occurs due to the inability of corrosion inhibitors getting to the top of the pipeline (12 o'clock) thereby exposing it to corrodents. The inhibition effect is found to be predominant at the bottom of the line (6 o'clock), 9 o'clock and 3 o'clock where the flow of the oil or gas is taking place. This exposes the top of the line to concerted attack by the agents of corrosion with a resultant failure at some point. The primary factor that affects TLC is temperature which acts on the iron carbonate film formed. The combined effect of temperature fluctuation and condensation rate exposes the iron carbonate film to deterioration and consequently more corrosion. A study of the influence of gas flow rate on TLC shows that higher flow rate (which results in higher condensation rate) brings about more corrosion [40], while at a certain critical condensation rate, temperature and pH, TLC does not occur in gas pipelines [41]. The presence of acetic acid (HAc) has been found to enhance CO_2 TLC on carbon steel pipe, though at certain concentration level, HAc does not affect CO_2 TLC in carbon steel [42].

External Corrosion

External corrosion is caused by water penetrating the insulation system and is trapped between the insulation and the external pipe wall. The corrosion cell is fuelled by the continual supply of water and oxygen from the external sources. The main area where external corrosion is found is at the field applied weld insulation packs, but it can also be at any location where the galvanized insulation jacket has been punctured or torn. Weld pack insulations that are not well sealed allow water ingress making the weld packs to be wet. A fairly high temperature is needed to drive the corrosion mechanism, and the longer the mechanism has been active, the worst the damage will be. Therefore, the hottest and coldest lines in the field should have the highest likelihood for having an external corrosion problem.

CORROSION MANAGEMENT TECHNIQUES

Corrosion management is that part of the overall management system, which is concerned with the development, implementation, review, and maintenance of corrosion policy [7]. The corrosion policy, however, is a framework on which decision concerning corrosion issue in an industrial setting is based. This framework provides basic measures for risk determination via development of absolute risk control measures through planning, implementation, and control strategies. Corrosion management contributes to numerous benefits like statutory or corporate compliance with safety, health and environment policies, reduction in leaks, increased plant availability, reduced unplanned maintenance, and reduction in deferment costs [43]. To manage corrosion involves the utilization of a framework that will model the organization's policy through organizing, planning and implementing, measuring and reviewing, and auditing performance at all levels of execution as shown in Figure 6.

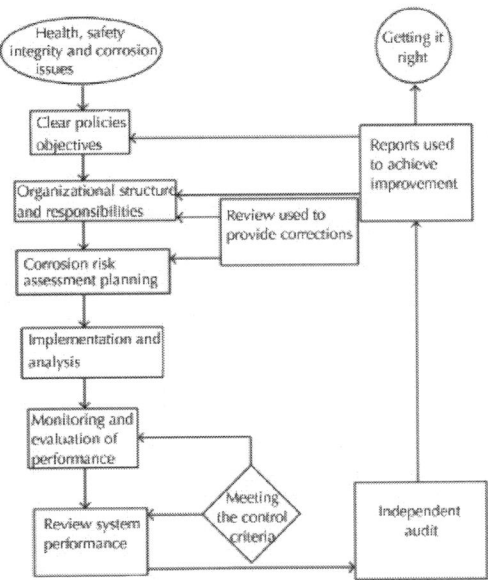

Figure 6: Corrossion management framework.

Corrosion Risk Assessment (CRA)

In planning for corrosion management, there is need for a formal identification of facilities that have the risk of degradation due to corrosion. The purpose of corrosion risk assessment is to rank facilities in order of their proneness to corrosion, identify options to remove, mitigate, or manage the risks. In order to manage corrosion risks, monitoring and inspection program will be incorporated in the overall activity schedule of an organization. The probability of failure is estimated based on the type of corrosion damage expected to occur on the component while the consequences of failure are measured against the impact of such a failure evaluated against a number of criteria. The criteria could include potential hazards to environment, risks associated with safety and integrity, or risk due to corrosion or inadequate corrosion mitigation procedure.

Typical of the risk-based procedure is the Failure Mode, Effect and Critical Analysis (FMECA) that ranks perceived risks in order of seriousness as shown in (18):

Criticality (Risk) = Effect (Consequences) * Mode (Probable Frequency),   (18)

Where: failure Criticality is potential failures as examined in order to predict the severity of each failure effect in terms of safety, decreased performance, total loss of function and environmental hazards. Failure effect is potential failures assessed to determine the probable effect on process performance and the effects of the components on each other. Failure mode is the anticipated operational conditions used to identify most probable failure mode, the damage mechanism and likely locations.

Corrosion risk is the product of the probabilities of a corrosion-related failure and the consequences of such a failure [44]. The risk analysis of a pipeline is a measure of the probability of failure. The acceptable annual failure probability is dependent on the safety class [45] as shown in Table 1.

Table 1: Safety class and target annual failure probability

Safety class	Annual failure probability
High	$<10^{-3}$
Medium	$<10^{-4}$
Low	$<10^{-5}$

Corrosion risk assessment can be carried out on a group of components which are constructed from the same material and subject to the same operating condition or an individual component. In oil and gas pipelines, the risk is analysed as either external or internal corrosion threat or environmental and operational threat. The remaining life of the pipeline is estimated against some established operational standards, while the rate of corrosion is correlated with the operating parameters of the oil and gas like CO_2, H_2S, temperature, pressure, flow rate, water cut and so forth. For effective corrosion assessment, the information concerning the operating condition of a facility will be maintained throughout the life cycle. The information is useful in formulating a corrosion risk assessment model that will be validated and modified with new assumptions overtime. For a non stable process condition, detailed re-assessment will be required at least annually but a stable process with good historical data trend will need revalidation less frequently [7].

Risk Based Inspection (RBI)

In managing oil and gas pipelines against corrosion, RBI technique is used to develop an optimum plan for the execution of the inspection activities. RBI uses findings from corrosion risk assessment (CRA) or other risk analysis to plan physical inspection procedures. A risk-based approach to inspection planning will ensure that risk is reduced to as low as reasonably practicable. It will also optimize inspection schedule, focus effort on the most critical area, and identify the most appropriate methods of inspection [46]. Planning a risk-based analysis involves listing activities, task and other elements of a project, identifying the technical risks, develop a risk ranking factor scale for each activity, document results and identify potential risks reduction actions for evaluation by personnel [47].

Corrosion Monitoring

Corrosion inspection and monitoring are key activities in ensuring, pipelines integrity are maintained and corrosion mitigated [48]. The choice of corrosion control measure is a function of fluid composition, pressure, temperature, aqueous fluid corrosivity, facility, and technical culture inherent in an establishment. In monitoring and inspection of pipelines, data are collected to enhance corrosion control by way of predicting the remaining life and the suggestion of possible mitigation measures that will help to enhance serviceability will largely depend on the experience of the personnel. A thorough practice for corrosion management involves the monitoring of corrosion risks through proactive and reactive monitoring techniques. In management of pipeline corrosion in oil and gas industries, proactive technique which involves determination of the corrosion standpoint prior to failure is utilized. This involves in-line and on-line monitoring system. In this system, data which could enhance the knowledge of the rate of corrosion degradation are collected and steps are taken to prevent failure. In-line system cover the installation of devices directly into the pipeline like corrosion coupons, biostuds and so forth. These need to be extracted for analysis periodically. On-line monitoring techniques include deployment of corrosion monitoring devices either directly into the process or fixed permanently to the facility. These include electrical resistance (ER) probes, linear polarization resistance (LPR) probes, fixed ultrasonic (UT) probes, acoustic emission and so forth.

Whereas some corrosion monitoring techniques can be used for continuous monitoring, others are used for periodic monitoring. Corrosion monitoring techniques can either be direct or indirect parameter measure. This is summarized in Table 2.

Table 2: Summary of corrosion monitoring techniques

Direct method	Indirect method
Non-destructive inspection (NDI)	Biological counts
Material test coupons	Hydrogen probes
Electrical resistance (ER) probes	pH probes
Linear polarization resistance (LPR)	Specific ions
Elector-chemical impedance spectroscopy (EIS)	Temperature
Electro-chemical noise (EN)	conductivity
Galvanic current (GC)	Electrical potential monitor

Corrosion Mitigation Strategies

After corrosion risk assessment and data collection and analysis are completed, there is need for corrective action on the facility; this depends on the level of the deterioration experienced by facility. The approaches available for mitigating corrosion in pipeline includes, coating surfaces to act as a barrier or perhaps provide sacrificial protection, the addition of chemical specie to the environment to limit corrosion, alteration of alloy chemistry to make it more resistance to corrosion and utilization of alternative material [24].

Effective corrosion mitigation involves a good approach to assessment linked to inspection monitoring during initial design and re-evaluation of pipeline with respect to the selection of inhibitors. The summary of inhibitor selection for carbon steel pipeline at different risk categories is shown in Table 4.

Corrosion can be prevented or controlled by understanding the principle underlying corrosion process. This understanding has been the basis for the development of a number of corrosion prevention measures. The basic corrosion control measures are based on electrochemical driving force as shown in Pourbaix diagram in Figure7. Table 3 shows the different pipeline corrosion mitigation strategies.

Table 3: Shows the different pipeline corrosion mitigation strategies

Mitigation strategy	Option	Remarks
Appropriate materials	Use of corrosion resistant alloys, non-metallic materials like Reinforced composite, thermoplastic-lined and polyethylene pipelines. Consider use of internally coated carbon steel pipeline systems (i.e., nylon or epoxy coated) with an engineered joining system.	(i) Non-metallic materials may be used as a liner or a free standing pipeline depending on the service conditions. (ii) Selection of appropriate material at construction and major refurbishment stage is necessary.
Chemical treatment	Corrosion inhibitors, biocides, oxygen scavengers, gas blanketing, vacuum deaeration	(i) The presence of small amounts of oxygen (parts per billion) or bacteria will accelerate corrosion. (ii) Provides a barrier between corrosive elements and the pipe surface
Coating and lining	Organic Coatings, metallic coatings, lining, cladding	Useful for internal and external corrosion prevention
Cathodic protection	Sacrificial anodes, impressed current systems, hybrid system	Need ability to monitor performance on-line.
Process control	Identify key parameters: pH, temperature, pressure, Flow rate, water chemistry, pH, chlorides, dissolved metals, bacteria, suspended solids, chlorine, oxygen, and chemical residuals	(i) Changes in operating conditions will influence the corrosion potential. Production information can be used to assess corrosion susceptibility based on fluid velocity and corrosivity (ii) Trends in dissolved metal concentration (i.e., Fe, Mn) can indicate changes in corrosion activity

Design detailing	Ensure ease of access and replacement: (i) Install valves that allow for effective isolation of pipeline segments from the rest of the system (ii) Install binds for effective isolation of in-active pipeline segments	Allows the effective suspension and discontinuation of pipeline segments: (i) Removes potential "deadlegs" from the gathering system (ii) Develop shut-in guidelines for the timing of required steps to isolate and lay up pipelines in each system

Table 4: Corrosion inhibitor risk categories

Risk category	Max inhibitor availability	Max expected uninhibited corrosion rate (mm/yr)	Comments	Proposed category name
1	0%	0.4	Benign Fluid, corrosion inhibitor use not anticipated. Predicted metal loss accommodated by corrosion allowance	Benign
2	50%	0.7	Corrosion inhibitor probably required but with expected corrosion rates there will time be time to review the need for inhibition based on inspection data.	Low
3	90%	3	Corrosion inhibition required for majority of field life but inhibitor facilities need not be available from day one.	Medium
4	95%	6	High reliance on inhibition for operational life time. Inhibitor facilities most be available from day one to ensure success	High
5	>95%	>6	Carbon steel and inhibition is unlikely to provide integrity for full field life. Select corrosion resistant material or plan for repair and replacement	Unacceptable

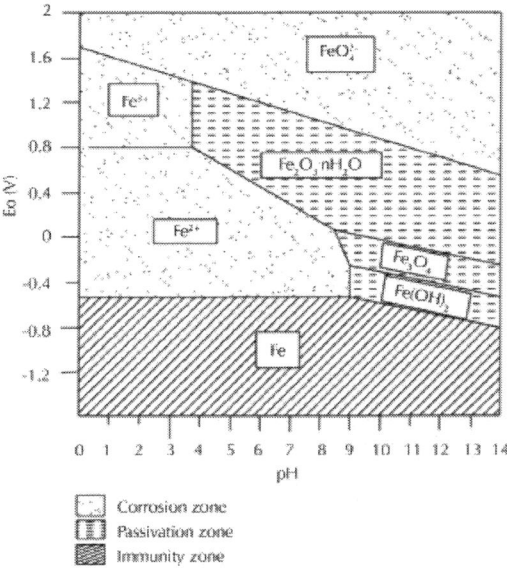

Figure 7: Pourbaix diagram of iron corrosion, passivity and immunity.

CONCLUSIONS

The prevalence of corrosion in oil and gas industry has resulted in enormous investment in technology to help combat impacts like loss of containment, leakages, death of personnel an environmental pollution. To this end, new oil and gas fields are developed using experiences generated from previous fields with similar characteristics. Efforts of design personnel at ensuring that the carbon steel materials are operating in the environment of immunity or passivity as shown in Pourbaix diagram (Figure 7) are yielding results via introduction of high corrosion resistant carbon alloys for pipelines. In other instances, specialized corrosion resistant materials have been used for lining the pipelines while reinforced composite and PVC materials have been utilized as alternative material for pipeline construction.

Advancements in inspection and monitoring techniques are also aiding corrosion experts in decision concerning the "when" and "how" pipelines are managed in a bid to optimize performance and cost.

The proliferation of different empirical, statistical, and mechanistic prediction models for corrosion prediction is aiding personnel in managing the integrity of the pipelines through different mitigation strategies.

Finally, if pipeline corrosion which is a major contributor to nonproductive time (NPT) in oil and gas production will be reduced to the barest minimum, a corrosion management policy with a well-defined structure that includes responsibilities, reporting routes, practices, procedures, and resources has to be strictly followed in the oil and gas industries. The effectiveness of the policy will therefore depend on the willing of the leadership and commitment of other personnel at all ranks.

REFERENCES

1.   R. Nicholson, J. Feblowitz, C. Madden, and R. Bigliani, "The Role of Predictive Analytics in Asset Optimization for the Oil and Gas Industry-White Paper," 2010, http://www.tessella.com/wp-content/uploads/2008/02/IDCWP31SA4Web.pdf.

2.   J. Neelamkavil, "A review of Existing Tools and Their Applicability to Maintenance Management. Report # RR-285,"http://pdf.aminer.org/000/274/575/a_decision_support_system_for_transmission_facility_maintenance.pdf.

3.   Oil and Gas Enhanced Production Services Industry to 2016—Enhanced Oil Recovery (EOR) Driving E&P Activity in Depleting Hydrocarbon Reservoirs, http://www.reportlinker.com/p0845623/Oil-and-Gas-Enhanced-Production-Services-Industry-to-Enhanced-Oil-Recovery-EOR-Driving-E-P-Activity-in-Depleting-Hydrocarbon-Reservoirs.html.

4.   Control of Major Accident Hazards, "Ageing Plant Operational Delivery   Guide,"http://www.hse.gov.uk/comah/guidance/ageing-plant-core.pdf.

5.   P. Horrocks, D. Mansfield, K. Parker, J. Thomson, T. Atkinson, and J. Worsley, "Managing Ageing Plant," http://www.hse.gov.uk/research/rrpdf/rr823-summary-guide.pdf.

6.   "Cost of Corrosion to Exceed $1 Trillion in the United States in 2012—G2MT Labs -The Future of Materials Condition

Assessment," http://www.g2mtlabs.com/2011/06/nace-cost-of-corrosion-study-update/.

7. Review of Corrosion Management for Offshore Oil and Gas Processing, HSE OffshoreTechnology Report 2001/044, 2001.

8. Corrosion in the Oil Industry (Oilfield review) Schlumberger,http://www.slb.com/resources/publications/industry_articles/oilfield_review/1994/or19940401_corrosion.aspx.

9. B. Khajota, D. Sormaz, and S. Nesic, "Case-based reasoning model of CO_2 corrosion based on field data," CORROSION, 2007, paper no. 07553.

10. K. U. Raju, "Successful scale mitigation strategies in Saudi Arabian oil fields," in International Symposium on Oilfield Chemistry, The Woodlands, Tex, USA, April 2009, paper no. 121679.

11. A. Darwin, K. Annadorai, and K. Heidersbach, "Prevention of corrosion in carbon steel pipeline containing hydrostatic water-an overview," in CORROSION, March 2010, paper no. 10401.

12. J. Wen, T. Gu, and S. Nesic, "Investigation of the effects of fluid flow on SRB biofilm," in CORROSION, 2007, Paper no. 07516.

13. B. Hedges, H. J. Chen, T. H. Bieri, and K. Sprague, "A review of monitoring and inspection technique for CO_2 and H_2S corrosion in oil and gas production facilities: location, location, location," in CORROSION, 2006, paper no. 06120.

14. M. Singer, B. Brown, A. Camacho, and S. Nešić, "Combined effect of carbon dioxide, hydrogen sulfide, and acetic acid on bottom-of-the-line corrosion," Corrosion, vol. 67, no. 1, 2011. ·

15. K. L. J. Lee and S. Nesic, "EIS investigation of CO_2/H_2S corrosion," in CORROSION, April 2004, paper no. 04728.

16. K. D. Ralston and N. Birbilis, "Effect of grain size on corrosion: a review," Corrosion, vol. 66, no. 7, pp. 0750051–07500513, 2010. ·

17. Y. Song, A. Palencsár, G. Svenningsen, J. Kvarekvål, and T. Hemmingsen, "Effect of O_2 and temperature on sour corrosion," Corrosion, vol. 68, no. 7, pp. 662–671, 2012.

18. A. Kale, B. H. Thacker, N. Sridhar, and C. J. Waldhart, "A probabilistic model for internal corrosion of gas pipelines," in Proceedings of the 5th Biennial International Pipeline Conference (IPC '04), pp. 2437–2445, Calgary, Canada, October 2004.

19. S. Nesic, J. Cai, and K.-L. J. Lee, "A multiphase flow and internal corrosion prediction model for mild steel pipeline," in CORROSION, 2005, Paper no. 05556.

20. W. Sun and S. Nesic, "A mechanistic model of H_2S corrosion of mild steel," in CORROSION, 2007, paper no. 07655.

21. X. Hu, V. D. Souza, A. Neville, and J. Well, "Prediction of erosion-corrosion in oil and gas- a systematic approach," in CORROSION, 2008, paper no. 08540.

22. X. Tang, C. Li, F. Ayello, J. Cai, and S. Nesic, "Effects of oil type on phase wetting transition and corrosion in oil-water flow," in CORROSION, NACE International, 2007, Paper no. 017170.

23. Y. Xian and S. Nesic, "A stochastic prediction model of localized CO_2 corrosion," in CORROSION, 2005, paper no. 05057.

24. CAPP, "Best Management Practices: Mitigation of Internal Corrosion in Oil Effluent Pipeline Systems," 2009, http://www.capp.ca/getdoc.aspx?DocId=155641&DT=PDF.

25. B. A. Shaw and R. G. Kelly, "What is corrosion?" Electrochemical Society Interface, vol. 15, no. 1, pp. 24–26, 2006.

26. J. Kruger, "Electrochemistry of Corrosion," 2001, http://electrochem.cwru.edu/encycl/art-c02-corrosion.htm.

27. V. Fajardo, C. Canto, B. Brown, and S. Nesic, "Effect of organic acids in CO_2 corrosion," in Proceedings of the NACE International Conference and Exposition CORROSION, 2007, paper no. 07319.

28. P. S. Joshi, G. Venkateswaran, K. S. Venkateswarlu, and K. A. Rao, "Stimulated decomposition of $Fe(OH)_2$ in the presence of AVT chemicals and metallic surfaces—relevance to low-temperature feedwater line corrosion," CORROSION, vol. 49, no. 4, pp. 300–309, 1993.

29. R. Nyborg, "Controlling internal corrosion in oil and gas pipeline," Business Briefing-Exploration & Production: The Oil & Gas Review, no. 2, pp. 70–74, 2005.

30. A. Dugstad, E. Gulbrandsen, M. Seiersten, J. Kvarekval, and R. Nyborg, "Corrosion testing in multiphase flow, challenges and limitations," in CORROSION, 2006, paper no 06598.

31. R. N. Kig, "A review of fatigue crack growth in air and seawater," Offshore Technology Report OTH96 511, HSE, 1996.

32. A. Keating and S. Nesic, "Prediction of two-phase erosion-corrosion in bends," in Proceedings of the 2nd International Conference on CFD in Minerals and Processes Industries CSIRO, Melbourne, Australia, December 1999.

33. S. Nesic and J. Postlethwaite, "Relationship between the structure of disturbed flow and erosion-corrosion," Corrosion, vol. 46, no. 11, pp. 874–880, 1990.

34. H. Wang, W. Paul Jepson, J.-Y. Cai, and M. Gopal, "Effect of bubbles on mass transfer in multiphase flow," in CORROSION, 2000, paper no. 00050.

35. H. Fang, B. Brown, and S. Nešiæ, "Effects of sodium chloride concentration on mild steel corrosion in slightly sour environments," in CORROSION, vol. 67, no. 1, January 2011.

36. E. Mysara Mohyaldinn, N. Elkhatib, and C. Mokhtar Ismail, "A computational tool for erosion/corrosion prediction in Oil/Gas production facilities," in Proceedings of 3rd International Conference on Solid State Science & Technology (ICSSST '10), Kuching, Malaysia, December 2010.

37. Sh. Hassani, K. P. Roberts, S. A. Shirazi, J. R. Shadley, E. F. Rybicki, and C. Joia, "Flow loop study of NaCl concentration effect on erosion, corrosion, and erosion-corrosion of carbon steel in CO_2-saturated systems," in CORROSION, vol. 68, no. 2, February 2012.

38. A. A. Sami and A. A. Mohammed, "Study synergy effect on erosion-corrosion in oil and gas pipelines,"Engineering and Technology, vol. 26, no. 9, 2008.

39. M. Baker Jr., "Stress Corrosion Cracking Study," 2004,http://www.polyguardproducts.com/products/pipeline/TechReference/SCC_Report-Final_Report_with_Database.pdf.

40. S. Olsen and A. Dugstad, "Corrosion under dewing conditions," in CORROSION, 1991, paper no. 472.

41. F. Vista and K. Alam, "Semi-empirical model for prediction of top-of-the-line corrosion risk," inCORROSION, 2002, paper no. 02245.

42. C. Mendex, M. Singer, A. Camacho, S. Hernndez, and S. Nesic, "Effect of acetic acid pH and MEG on CO_2 top of the line corrosion," in CORROSION, 2005, paper no. 05278.

43. D. Storey, "A Service Company's Experience with Pipeline Integrity Management," 2004, http://www.roseninspection.net/MA/papers/2004-11_PipelineIntegrityManagement.pdf.

44. P. O. Gartland and J. Roy, "Application of internal corrosion modelling in risk assessment of pipeline," inCORROSION, 2003, paper no. 03179.

45. Det Norske Veritas (DNV RPG 101), Recommended Practice DNV-RP-101: Corroded Pipelines, 2010.

46. Det Norske Veritas (DNV RPG 101), "Risk Based Inspection of Topsides Static Mechanical Equipment, 2001".

47. P. K. John and L. D. John, "Risk factor analysis—a new qualitative risk management tool," in Proceedings of the Project Management Institute Annual Seminar & Symposium, September 2000.

48. E. J. Carl, A. B. John, and G. T. Neil, "Improving plant reliability through corrosion monitoring," inProceedings of the Process Plant Reliability, Houston, Tex, USA, November 1995.

Chapter 2

Minimising Hydrogen Sulphide Generation During Steam Assisted Production of Heavy Oil

Wren Montgomery[1], Mark A. Sephton[1], Jonathan S. Watson[1], Huang Zeng[2], and Andrew C. Rees[3]

[1]Department of Earth Science and Engineering, Imperial College London, SW7 2AZ, UK, 2

[2]Heavy Oil Technology Flagship, BP Canada Energy Group ULC, 240 4th Ave SW, Calgary T2G 0N3, Canada, 3

[3]Heavy Oil Technology Flagship, BP America Production Co, 501 Westlake Park Boulevard, Houston, TX 77079, USA

ABSTRACT

The majority of global petroleum is in the form of highly viscous heavy oil. Traditionally heavy oil in sands at shallow depths is accessed by large scale mining activities. Recently steam has been used to allow heavy oil extraction with greatly reduced surface disturbance. However, in situ thermal recovery processes can generate hydrogen sulphide, high levels of which are toxic to humans and corrosive to equipment. Avoiding hydrogen sulphide production is the best possible mitigation strategy. Here we use laboratory aquathermolysis to reproduce conditions that may be experienced during thermal extraction. The results indicate that hydrogen sulphide generation occurs within a specific temperature and pressure window and corresponds to chemical and physical changes in the oil. Asphaltenes are identified as the major source of sulphur. Our findings reveal that for high sulphur heavy oils, the generation of hydrogen sulphide during steam assisted thermal recovery is minimal if temperature and pressure are maintained within specific criteria. This strict pressure and temperature dependence of hydrogen sulphide release can allow access to the world's most voluminous oil deposits without generating excessive amounts of this unwanted gas product.

INTRODUCTION

Heavy oil represents the majority of petroleum on Earth with seven trillion barrels of resources in place. Major reserves of heavy oil exist within deposits of unconsolidated sand and clay. For example, the oil sands of Alberta, Canada, represent a staggering resource of 1.8 trillion barrels[1]. Initially, those reservoirs with a depth of less than 70 meters were accessed by straightforward mining[2]. Oil sands at depths greater than 70 meters cannot be economically developed by surface mining, and other recovery methods are needed. Using Alberta, Canada, as an example once more, 750 million barrels were recovered by steam assisted in situ methods in 2012 and production from in situ methods is expected to double by 2022[1]. The most popular variant of thermal in situ recovery is steam assisted gravity drainage[3] which uses an injection well to produce a steam saturated zone. The oil is heated and its viscosity reduced to promote flow. A second, lower, producer well collects the mobilized oil which is then pumped to the surface.

Gas generation due to thermally driven reactions is a recognized challenge during steam assisted in situ recovery of high sulphur heavy oils including that in the Alberta oil sands. Sulphur-rich organic matter is notoriously chemically reactive[4] and can lead to the release of hydrogen sulphide following treatment with steam. Many heavy oils are tainted by high sulphur contents (>2 w%[5]) owing to the concentration and generation of organic sulphur compounds during biodegradation[6]. Capture of the hydrogen sulphide produced during steam recovery is necessary. Owing to the large economic penalties of overestimating or underestimating sulphur production, accurate appraisals increase operational efficiency.

Here we investigate the chemistry, temperature and pressure dependence on gas production occurring during steam assisted recovery. We have used laboratory aquathermolysis to simulate subsurface conditions and monitor the change in product composition. A quantitative approach, with recovery and analysis of gas, water soluble organic compounds, liquid oil, asphaltene and mineral matrix adsorbed residue, allows the recognition of material transfer between solubility and volatility fractions.

RESULTS

Our aquathermolysis experimental products were comprehensively analysed using preparative chromatography, FTIR spectroscopy, gas chromatography-flame photometric detection, gas chromatography-mass spectrometry and pyrolysis-gas chromatography-mass spectrometry. Our data correspond to the following temperature and pressure intervals: 150°C (<690 kPa; <100 psig), 175°C (<690 kPa; <100 psig), 200°C (1030 kPa; 150 psig), 225°C (2760 kPa; 400 psig), 250°C (4140 kPa; 600 psig), 275°C (5520 kPa; 800 psig), 300°C (8620 kPa; 1250 psig) and 325°C (12070 kPa; 1750 psig). Each temperature and pressure combination represents a unique experiment run for 24 hours. The duration was chosen so as to avoid the short term (<12 hrs) kinetic effects which have been previously documented for similar oil sand samples[7, 8, 9] and to access the steady state chemistry of the system. Experimental pressure is a result of the volume increase of water and other fluids present and is read from a dedicated pressure gauge. For simplicity each experiment will be referred to hereafter using only its temperature. Our study uses a sulphur-rich oil sand from Canada as a typically problematic resource. The Alberta oil sands deposit in western Canada contains over 1.8 trillion barrels of heavy oil in Lower Cretaceous Mannville Group sands at typical depths of

<100–500 m. In situ thermal methods are particularly appropriate for the Canadian oil sands because 80% are too deep to be accessed by mining[7]. The sulphur-rich nature of the oil suggests a source dominated by the Carboniferous Exshaw Formation in the eastern portion of the deposit and Jurassic Gordondale Member in the west[10, 11]. Our oil sand came from the north east of the region and contained 18.5% extractable organic matter. The solvent-extracted oil had the following characteristics: $C = 82.5\%$; $H = 10.0\%$; $N = 0.4\%$; $S = 5.34\%$; saturates $= 29.6\%$; aromatics $= 40.5\%$; resins $= 10.6\%$; asphaltenes $= 19.3\%$.

Physical and Chemical Fractions

Our data set (Figure 1) allows us to define a number of physical products observable following aquathermolysis experiments. Gas is generated in the reactor headspace, flotate is a light oil that floats on the surface of the water, sinkate is a heavier oil that sinks in the water and residue is an oil that coats mineral surfaces. The relative abundances in the three liquid products in each of the experiments are displayed in Figure 1a.

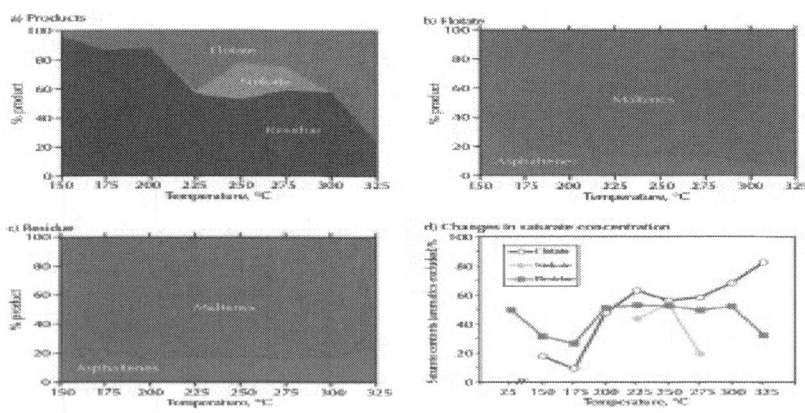

Figure 1: (a) relative proportions of flotate, sinkate and residue, (b) relative proportions of maltenes and asphaltenes in the flotate, (c) relative proportions of maltenes and asphaltenes in the residue, (d) Changes in saturate concentration for the experimental products expressed as saturates/(saturates + resins + asphaltenes) × 100. Aromatic hydrocarbons change little during aquathermolysis, consistent with published work[12], and so have been excluded.

There is evidence that the physical properties of the products are controlled by their chemical constitution. Each of the recovered products can be separated into the relatively high molecular weight asphaltenes and lower molecular weight maltenes (Figure 1b, c). Division of the maltenes into saturate, aromatic and resin fractions (Figure 1d) completes "SARA" analysis. The relative abundances of SARA fractions within the flotates, sinkates and residues are different and vary with changing temperature and pressure of aquathermolysis. Some patterns exist and aromatic hydrocarbons change little during aquathermolysis[12] and can be considered relatively constant: resins and asphaltenes display similar trends with temperature and pressure and can be summed into a single fraction. Chemical variations can, therefore, be expressed simply as saturates/(saturates + resins + asphaltenes) × 100 (Figure 1d).

The flotate first appears in the 150°C experiment and its abundance then increases with experimental temperature (Figure 1a). The chemistry of the flotate changes with temperature and its asphaltene content decreases (Figure 1b) to produce an increasingly lighter and less viscous oil. Aquathermolysis treatment up to 325°C progressively reduces viscosity in the flotate from 1105 Pa*s to 1.5 Pa*s. Examination of the chemical fractions within the flotate reveals that the flotate maltenes become more saturate rich with temperature and pressure (Figure 1b) implying that hydrocarbons are being introduced from degradation of the functionalised resin or high molecular weight asphaltene structures.

A sinkate joins the flotate in experiments carried out above 225°C (Figure 1a) and a divergence in chemical composition begins, with the flotate becoming more saturate rich and the sinkate becoming more resin and asphaltene rich (Figure 1d). The chemical dichotomy which appears to maintain the two phases disappears above 275°C. The saturated compounds appearing in the flotate must be derived from a reservoir within the resins and asphaltenes which becomes exhausted above 275°C. In the context of in situ steam assisted production the appearance of a two-phase liquid between 225°C and 300°C driven by thermally induced chemical changes in the oil may affect fluid flow characteristics in the reservoir with each phase being capable of movement at different rates.

Our chemical composition data reveal that there is a notable difference in susceptibility to chemical change for oil coating mineral surfaces relative to that free in the experimental products. While the chemical composition of the flotate and sinkate begins to vary in experiments above 225°C, the residue appears to resist chemical change (Figure 1 c, d). Between 200°C and 300°C the relative abundance of maltenes relative to asphaltenes and the saturates/(saturates + resins + asphaltenes) × 100 values for the residue are relatively constant. In terms of heat energy, the protective effects of mineral scaffolding equates to a differential of 75°C, highlighting the importance of geological and mineralogical context in oil sand reservoirs.

Hydrogen Sulphide Production

Aquathermolysis also leads to significant changes in headspace gas composition. Hydrogen sulphide is quantifiable by gas chromatography-flame photometric detection at experimental temperatures above 200°C. Notably the onset of hydrogen sulphide generation directly precedes the temperature at which a loss of resins and asphaltenes begins and at which there is a divergence in the saturate contents of the flotate and sinkate (Figures 1d & 2). The correlation implies an origin for hydrogen sulphide by the degradation of sulphur-containing units in the resin or asphaltene fractions.

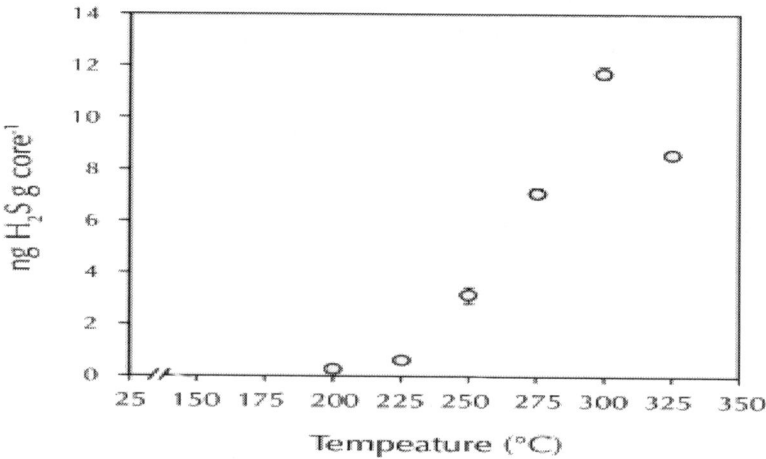

Figure 2: Hydrogen sulphide produced by aquathermolysis of sulphur rich heavy oil.

300 and 325°C data points represent minimum values. Error bars are generated from the variation in multiple measurements on the gas produced from a single experiment.

Comprehensive analysis of headspace gas is achieved by gas chromatography-mass spectrometry (Figures 2 and 3a). Hydrogen sulphide increases markedly with experimental temperatures above 200°C and is accompanied by the generation of methane and C_2 to C_6 saturated hydrocarbons (ethane, propane, butane and hexane). The presence of C_2 to C_6 saturated hydrocarbons is commonly interpreted as evidence of the thermal degradation of higher molecular weight organic matter, supporting the proposal that hydrogen sulphide has an origin by the thermal dissociation of an organic fraction. At experimental temperatures at and above 300°C the responses of C_5 and C_6 diminish and the responses of shorter chain saturated hydrocarbons (C_1 and C_2) are enhanced, features which are in harmony with a greater number of bond breaking events occurring at higher temperatures.

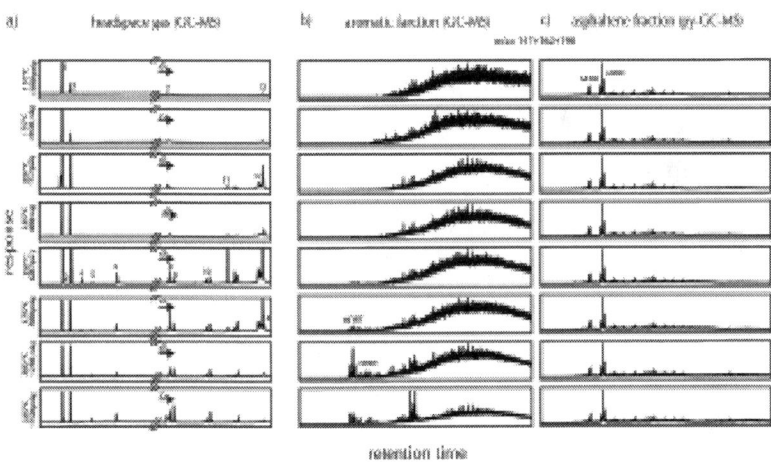

Figure 3: (a) Total ion current of gas species generated by aquathermolysis of sulphur-rich heavy oil. (1) argon(from purge), (2) methane, (3) carbon dioxide, (4) ethylene, (5) ethane, (6) hydrogen sulphide, (7) water, (8) propene, (9) propane, (10) butene isomers, butane, (11) dichloromethane (DCM contamination), (12) pentane, (13) hexane, (14) benzene, alongside a comparison of selected ions (m/z = 147 + 162 + 198) from the (b) aromatic (GC-MS) and (c) asphaltene (py-GC-MS) fractions of the flotates. MDBT – methyldibenzothiophene, DDBT – dimethyldibenzothiophene.

A full appreciation of hydrogen sulphide generation during aquathermolysis requires identification of transformable sulphur in the starting materials. Inorganic sulphur is an unlikely source in an oil-stained sandstone, because sandstone rocks contain no sulphur-bearing minerals, leaving organic sources as the remaining possibility. The proposal that hydrogen sulphide is derived from thermal degradation of sulphur-containing organic units can be tested by subjecting the generated intermediate molecular weight hydrocarbon liquids to gas chromatography-mass spectrometry and the relatively high molecular weight asphaltenes to on-line pyrolysis gas chromatography-mass spectrometry.

Organic sulphur is absent in the saturate fractions, limited in the resin fractions but clearly present in the aromatic and asphaltene fractions. Data from the flotate aromatic fraction are displayed in Figure 3b and data from on-line pyrolysis of asphaltenes are presented in Figure 3c. A number of products are present in the total ion current which can be filtered using partially reconstructed ion chromatograms (m/z = 147 + 161 + 198) to reveal the presence of thiophenes, benzo- and dibenzo- thiophenes (Figure 4). Specifically, the aromatic fractions of the flotates and residues generated in experiments at 275°C and above contain various benzo- and dibenzothiophenes.

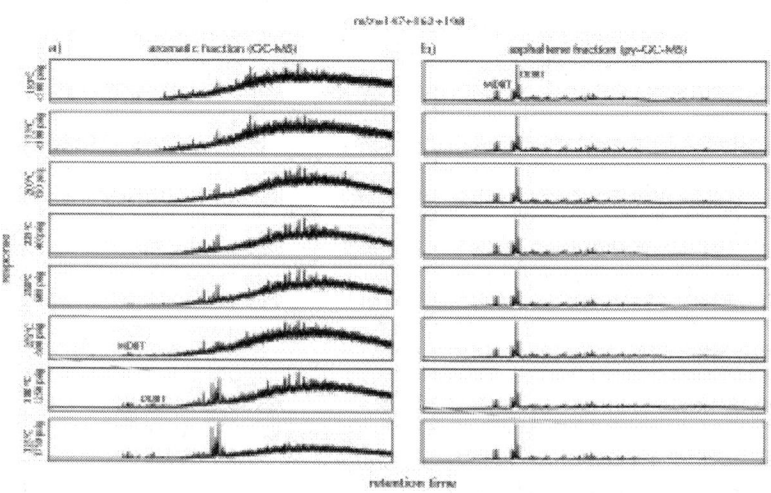

Figure 4: Comparison of selected ions from the (a) aromatic (GC-MS) and (b) asphaltene (py-GC-MS) fractions of the residues.

Equivalent data for the flotate fraction is presented in Figure 3b and 3c. MDBT – methyldibenzothiophene, DDBT – dimethyldibenzothiophene.

The asphaltene fractions of flotates and residues generated at all experimental temperatures contained a plethora of sulphur-bearing molecules including thiophenes, benzo- and dibenzo- thiophenes. The population of sulphur-bearing molecules in the asphaltene fraction did not vary greatly between flotate and residue, nor were there any trends observed in the size or composition of sulphur-bearing molecules with increasing experimental temperature. Full lists of sulphur compounds identified in the flash pyrolysates of flotates and resides of asphaltenes are presented in Supplementary Table S5.

Notably, while present in the asphaltene fraction of the flotate phase at all reaction temperatures, the sulphur-bearing compounds are present in the flotate phase only at temperatures at and above 275°C, coincident with the onset of peak hydrogen sulphide production observed in Figure 2. Hence it appears that sulphur-rich portions of the asphaltenes are degrading at and above 200°C to produce a relatively small amount of hydrogen sulphide gas; more extensive degradation at and above 275°C leads to the main phase of hydrogen sulphide release and the additional liberation of moderate molecular weight sulphur containing units which then pass into the liquid flotate.

DISCUSSION

As the quantitatively dominant phase degrading during hydrogen sulphide production the asphaltene fraction should exert overall control on headspace gas composition. Assessments of the ability of oils to produce deleterious sulphur gases should therefore focus on this relatively high molecular weight material. Pyrolysis gas chromatography-mass spectrometry of asphaltenes from a heavy oil will immediately indicate whether hydrogen sulphide generation at and above 200°C will occur.

The observation of physical separation into light saturate-rich and heavy asphaltene-rich oil phases and the relative unreactivity of the heavy residue suggests the possibility of in situ upgrading. The residue does not respond until the very highest temperature (325°C) and the saturate-rich flotate will be much more mobile that the relatively saturate-poor sinkate. The two free phases will flow at different rates allowing the selective recovery of lighter hydrocarbons that are more easy to handle and of higher economic value. There are also substantial economic benefits derived from leaving the less economic resins and asphaltenes behind in the reservoir related to the avoidance of extensive and expensive surface refining. Allowing a substantial part of the refining to occur in the reservoir also confers substantial environmental benefits. In the reservoir, the two phase oil generated by medium temperature (225°C to 300°C) aquathermolysis may be a characteristic of sulphur rich heavy oil which ensures the presence of resin and asphaltene structures that can be degraded at relatively low temperatures owing to the relative weakness of sulphur-containing bonds[4, 12].

Our data reveal that reservoir heterogeneity may also be an important factor. The main source of sulphur is the asphaltene fraction. Asphaltenes and associated organic sulphur compounds are enhanced following biodegradation and this process can be highly variable depending on access to microbes, water and oxidants and the pasteurizing effects of reservoir temperature. An example of variation in biodegradation levels is provided by the Canadian oil sands in which biodegradation increases to the east and north within the oil sands deposit and anti-correlates with maturity in the region[13].

The quantification of hydrogen sulphide allows predictions to be made for the amounts of this gas released during steam assisted production. Significant hydrogen sulphide generation begins at 200°C and amounts become substantial above 250°C. Our data are in accord with that of previous laboratory aquathermolysis studies[14] that defined 200°C for the onset of hydrogen sulphide generation. Field data display a less precise agreement[15] but may reflect less well constrained subsurface conditions and chemistries relative to laboratory experiments. Notably, 200°C is above that normally associated with steam assisted gravity drainage, although local hot spots occur. Other thermal methods such as cyclic steam simulation do breach the 200°C threshold. It must be noted that our hydrogen sulphide values are specific to the heavy oil used in the experiments. Substantial natural variation in sulphur contents are expected. Yet, although total yields may vary the basic principles of gas production with temperature are widely applicable and it is clear that hydrogen sulphide generation can be avoided if temperatures in the reservoir are not allowed to exceed 200°C. If minimizing hydrogen sulphide gas production is not a priority, the in situ upgrading characteristics of medium temperature (225°C to 300°C) aquathermolysis can be exploited. As the two phases observed have differing densities, it may be possible to extract the saturate-rich flotate while leaving the more sulphurous sinkate behind.

METHODS

Post Experiment Sampling

Once the experiment was complete and the reactor cooled, gases were extracted from the reactor headspace by releasing a needle gauge and allowing the gas to flow into a Tedlar collection bag with a septum. Gas was collected from this bag using a gas-tight syringe. The gas quantities in the syringe were scaled to reveal the total quantities in the reactor headspace. Following gas sampling, the reactor was opened and any oil floating on the surface of the water (the flotate) was transferred to a test tube using a polytetrafluoroethylene (PTFE)-coated spatula. To isolate any water-soluble organic matter, the water contents were pipetted from the vessel into test tubes, aided by rinsing with deionised water. If a layer of separated oil lay under the water but over the sediments, this was transferred to a test tube using a PTFE coated spatula. The remaining contents were transferred to the test tubes containing the water. These test tubes were centrifuged at 3000 rpm for 10 minutes to settle and combine any suspended fines with the solid residue, following which the supernatant was pipetted off. Two cycles of further addition of deionized water, centrifugation and pipetting followed. The separated water, containing the water-soluble fraction of the liquid water treatment products, was subjected to solid phase extraction (SPE). Then the solvent was allowed to evaporate to produce a water soluble oil fraction (the liquid-liquid extract). The residual oil still coating the sand was extracted with 93:7 dichloromethane/methanol (DCM/MeOH) by sonication for 15 minutes, followed by centrifugation and collection of the supernatant as described above. The remaining residual organic material was then extracted as above using toluene as the solvent. Any material remaining in the vessel is removed with 93:7 DCM/MeOH and any material resistant to that (cake) is removed using force.

Viscosity and Elemental Analysis

Rheological measurements were carried out using a Brookfield R/S Plus rheometer and analysed with RHEO3000 software. The flow curves were obtained by registering viscosity and shear stress at shear rates from 1 to 100 s^{-1}. Where possible, measurements were performed twice, at a temperature of 25°C ± 0.1. Determination of carbon, nitrogen and hydrogen was performed by the ASTM D5291 method where the sample is combusted in a pure oxygen environment and measured by Thermal Conductivity or IR. Measurement of sulphur was achieved by the ASTM D2622 method where the sample is placed in an X-ray beam and the peak intensity of the sulphur $K\alpha$ line at 5.373 A is determined.

Aquathermolysis

For the oil sand extraction experiments, sulphur-rich (%) oil in a sand matrix was obtained from north eastern of the Alberta oil fields. Approximately 50 g of oil sand were added to a high pressure and temperature reactor (Model 4740, Parr Instruments). Using the measured reactor volume, 75 ml, calculated sample density, and the pressure required, the mass of water needed was determined and added. The remaining headspace within the reactor was purged with argon and the reactor was sealed and placed in an oven for 24 hours. Pressure was monitored throughout the experiment using a dedicated pressure gauge attached to the reactor. Aquathermolysis experiments were performed on samples of oil sands at 25°C intervals from 150–325°C and corresponding pressures ranging from 100–2000 psi (Table 1). The experimental approach has been previously proved successful for heavy oil samples[16].

Table 1: Experimental temperatures and corresponding pressures

Temperature, °C	150	175	200	225	250	275	300	325
Pressure (kPa)	<690	<690	1030	2760	4140	5520	8620	12070
Pressure (psig)	<100	<100	150	400	600	800	1250	1750

Fractionation and Quantitative Analysis

Samples of the flotates, sinkates and 93:7 DCM/MeOH-soluble fractions were further separated into saturate, aromatic, resin (polar) and asphaltene (collectively termed SARA) fractions. Asphaltenes were precipitated from the sample using an excess of n-heptane with a 12 hour settling time; further separation was achieved by centrifugation. The precipitation process was repeated three times in total. The maltene fraction was further separated into saturate, aromatic, and polar fractions using alumina micro-column chromatography. The aliphatic fraction was eluted with hexane, the aromatic fraction with DCM, and the polar fraction with MeOH. Fractions were allowed to dry and were weighed.

Gas Chromatography-Flame Photometric Detector (FPD)

Gases produced were analysed using an Perkin Elmer AutoSystemXL gas chromatograph-flame photometric detector (GC-FPD). Samples (500 μl) were injected using a gas tight lockable syringe at a 10:1 split at with the injector at 220°C. Separation was performed on a J&W GS-Q column (30 m length, 0.32 mm internal diameter). Helium at column flow rate of 2 ml min^{-1} (constant flow rate) was used as the carrier gas. The GC oven temperature was held for 4 min at 35°C and then programmed at 10°C min^{-1} to 220°C, the final temperature was held for 6 min. The FPD was held at 300°C with a hydrogen flow of 75 ml min^{-1} and air 90 ml min^{-1}.

Gas Chromatography-Mass Spectroscopy and On-line Pyrolysis-Gas Chromatography-Mass Spectroscopy

Gases produced were analysed using an Agilent Technology 7890–5975 gas chromatograph-mass spectrometer (GC-MS). Samples (500 μl) were injected using a gas tight lockable syringe at a 5:1 split at with the injector at 220°C. Separation was performed on a J&W GS-Q column (30 m length, 0.32 mm internal diameter). Helium at column flow rate of 2 ml min^{-1} (constant flow rate) was used as the carrier gas. The GC oven temperature was held for 4 min at 35°C and then programmed at 10°C min-1 to 220°C, the final temperature was held for 6 min. The MS was run in full scan mode (10–200 a.m.u.) at 3.7 Hz. For on-line pyrolysis, asphaltene samples were placed in quartz sample tubes and loaded into a CDS Analytical Model 2500 Plus autosampler pyrolysis unit then subjected to pyrolysis (610°C at a heating rate of 20°C min^{-1}). The products were introduced to the GC via split injection at a 25:1 split ratio with the injector at 270°C. Separation was performed on a J&W DB-5 column (30 m length, 0.32 mm internal diameter). Helium at column flow rate of 2 ml min^{-1}(constant flow rate) was used as the carrier gas. The GC oven temperature was held for 2 min at 40°C and then programmed at 5°C min^{-1} to 310°C, the final temperature was held for 8 min. The MS was run in full scan mode (10–200 a.m.u.) at 3.7 Hz.

AUTHOR CONTRIBUTIONS

W.M. carried out the experiments, analysed the data and wrote the paper, M.A.S. designed the experiments, analysed the data and wrote the paper, J.S.W. designed the experimental equipment and analysed the data, H.Z. provided the samples, analysed the data and wrote the paper, A.C.R. designed the experiments, provided the samples, and wrote the paper. All authors discussed the results and implications and commented on the manuscript at all stages.

ACKNOWLEDGEMENTS

The authors are grateful to the STFC and UK Space Agency for a grant that enabled preliminary work for the project.

REFERENCES

1.   Energy Resource Conservation Board. *Alberta's energy reserves 2012 and supply/demand outlook 2013–2022.* (2013) Available at: http://www.aer.ca/documents/sts/ST98/ST98-2013.pdf

(Accessed: 3 May 2014).

2.   Schramm, L. L. *Surfactants: fundamentals and applications in the petroleum industry.*(Cambridge University Press, Cambridge, 2000).

3.   Butler, R. Application of SAGD, related processes growing in Canada. *Oil Gas J.* 99, 74–78(2001).

4.   Lewan, M. D. Sulphur-radical control on petroleum formation rates. *Nature* 391, 164–166(1998).

5.   Speight, J. G. *Enhanced recovery methods for heavy oil and tar sands.* (Gulf Publishing Company, Houston, USA, 2009).

6.   Méhay, S., Adam, P., Kowalewski, I. & Albrecht, P. Evaluating the sulphur isotopic composition of biodegraded petroleum: The case of the Western Canada Sedimentary Basin. *Org. Geochem.* 40, 531–545 (2009).

7.   Lamoureux-Var, V. & Lorant, F. H_2S artificial formation as a result of steam injection for EOR: a compositional kinetic approach. Paper presented at the SPE International Thermal Operations and Heavy Oil Symposium, Calgary, Alberta, Canada. doi:10.2118/97810-MS. (2005, November 1–3).

8.   Kapadia, P. R., Kallos, M. S. & Gates, I. D. A new kinetic model for pyrolysis of Athabasca bitumen, *Can. J. Chem. Eng.* 91, 889–901 (2013).

9.   Marcano, N., Larter, S., Snowdon, L. & Bennett, B. An overview of the origin, pathways and controls of H_2S production during thermal recovery operations of heavy and extra-heavy oil. Paper presented at GeoConvention 2013: Integration, Calgary, Alberta, Canada. Calgary: Canadian Society of Exploration Geophysicists. (2013, May 6–10).

10. Government of Alberta. *Responsible actions: a plan for Alberta's oil sands annual progress report 2010.* (2011) Available at:http://www.energy.alberta.ca/pdf/OSSResponsibleActionsProgressReport2011.pdf(Accessed: 3 May 2014.)

11. Adams, J., Marcano, N., Oldenburg, T., Mayer, B. & Larter, S. Sulphur and nitrogen compounds reveal oil sands source. Paper presented at GeoCanada 2010: Working with the Earth, Calgary,

Alberta, Canada. Calgary: Canadian Society of Petroleum Geologists. (2010, May10–14).

12.   Lewan, M. D. Laboratory simulation of petroleum formation by hydrous pyrolysis. *Organic Geochemistry* Engel, M.H. & Macko, S. A. (eds.) 419–422 Plenum Press, New York, (1993).

13.   Adams, J., Riediger, C., Fowler, M. & Larter, S. Thermal controls on biodegradation around the Peace River tar sands: Paleo-pasteurization to the west. *J. Geochem. Explor.* 89, 1–4(2006).

14.   Hyne, J. B., Clark, P. D., Clarke, R. A. & Koo, J. Aquathermolysis – the reaction of organosulphur compounds during steaming of heavy oils. Paper presented at the 2nd UNITAR International Conference on the Future of Heavy Crude and Tar Sands, Caracas, Venezuela. New York:McGraw-Hill. (1982, February 7–17).

15.   Thimm, H. F. Prediction of hydrogen sulphide production in SAGD, *J. Can. Petrol. Technol.*47, 7–9 (2008).

16.   Montgomery, W., Court, R. W., Rees, A. C. & Sephton, M. A. High temperature reactions of water with heavy oil and bitumen: Insights into aquathermolysis chemistry during steam-assisted recovery. *Fuel* 113, 426–434 (2013).

Enhancing Scientific Response in a Crisis: Evidence-based Approaches from Emergency Management in New Zealand

Emma E H Doyle[1], Douglas Paton[2], and David M Johnston[1, 3]

[1]Joint Centre for Disaster Research, Massey University, Wellington 6140, New Zealand

[2]School of Psychology, University of Tasmania, Newnham Campus, Locked Bag 1342, Launceston 7250, TAS, Australia

[3]GNS Science, Lower Hutt 5010, New Zealand

ABSTRACT

Contemporary approaches to multi-organisational response planning for the management of complex volcanic crises assume that identifying the types of expertise needed provides the foundation for effective response. We discuss why this is only one aspect, and present the social, psychological and organizational issues that need to be accommodated to realize the full benefits of multi-agency collaboration. We discuss the need to consider how organizational culture, inter-agency trust, mental models, information management and communication and decision making competencies and processes, need to be understood and accommodated in crisis management planning and delivery. This paper discusses how these issues can be reconciled within superordinate (overarching) management structures designed to accommodate multi-agency response that incorporates decision-making inputs from both the response management team and the science advisors. We review the science advisory processes within New Zealand (NZ), and discuss lessons learnt from research into the inter-organisational response to historical eruptions and exercises in NZ. We argue that team development training is essential and review the different types of training and exercising techniques (including cross training, positional rotation, scenario planning, collaborative exercises, and simulations) which can be used to develop a coordinated capability in multiagency teams. We argue that to truly enhance the science response, science agencies must learn from the emergency management sector and embark on exercise and simulation programs within their own organisations, rather than solely participating as external players in emergency management exercises. We thus propose a science-led tiered exercise program, with example exercise scenarios, which can be used to enhance both the internal science response and the interagency response to a national or international event, and provide direction for the effective writing and conduct of these exercises.

INTRODUCTION

During a volcanic crisis, whether an isolated period of unrest or a full scale eruption and recovery, many agencies and organisations are involved in its response and management. These range from expert and technical advisors (e.g., geologists, geophysicists, engineers, and social scientists), through to emergency management agencies (civil defence, fire service, police, army, national and local government) and lifeline organisations (lines companies, transport, water). For example, during the 1980 eruptions of Mt St Helens, over 130 officials and organisations responded (Saarinen and Sell [1985]); during the eruptive episodes of 1995 to 1996 in Ruapehu, NZ, over 42 organisations were involved (Paton et al. [1998a]), and in the 2012 Te Maari eruptions of NZ, some 30 organisations responded.

The number of responding agencies increases as the unrest or eruptive period continues, and they need to collaborate and share knowledge to effectively respond in a crisis that creates multiple, diverse consequences. However, these organizations bring to the crisis management context diverse operational roles, different organizational objectives and political or economic pressures, and varied ways of interpreting, prioritizing and responding to issues that reflect organizational policies, practices and cultures that range from emergency services' command and control practices to the more organic approach typical of scientific/research organizations. Ensuring that agency representatives can integrate their knowledge and expertise for response planning and implementation, and ensuring that they can continue to do so in a response environment that present complex, dynamic demands that need to be understood and managed over time, is a challenging task. We present here a literature review of the factors influencing response effectiveness, and discuss several approaches that have been developed to achieve an effective coordinated outcome, as well as how they could be integrated into the volcanology community to enhance and inform the response of volcanologists in the unique management environment created by volcanic crises. In the context of this literature, we also review and evaluate examples of NZ volcanic science advice and response practices.

First we discuss the development of science advisory groups in New Zealand since one of the earliest volcanic exercises run in 1992 (Section 2). We then discuss psychological, social and organizational influences and practices that affect response effectiveness (Section 3) and argue for the need for regular training activities to improve these competencies (Section 4). In doing so, we summarize the benefits accruing from science agencies learning from the methods that emergency management agencies routinely use, and embarking on exercise and simulation programs that mirror the complexities of the response environment in which they will make important, but non-routine, contributions within their own organisations. The paper argues that such in-house training (e.g., involving cross training, positional rotation, scenario planning, collaborative exercises, simulations, training and shared exercise writing tasks see Sections 4 and 5) is pivotal to developing the future response capability of science advisory groups and their ability to effectively complement the emergency management functions they will inevitably interact with. We also review national and international Civil Defence and Emergency Management (CDEM) exercises (Section 5) to highlight the benefits that can arise if scientists develop their own activities rather than solely participating as external players in emergency management exercises. Following a review of NZ's 2008 "All of nation" volcanic Exercise Ruaumoko (Section 5.2) to illustrate how effective evaluation informs the development of Volcanic Science Advisory Group (VSAG) processes within NZ, the paper concludes with our proposing, in Section 6, a new exercise structure for volcanology. This will facilitate the integration of volcanological expertise into CDEM processes and how scientists and scientific agencies can proactively contribute to and capitalise on opportunities to enhance shared understanding between diverse responding agencies. Throughout this paper, we consider science and science advice providers to represent the expert source of information on hazard processes (e.g. geology, geophysics, geochemistry, geodesy, atmospheric science) and the expert source of information on social and economic impacts, including communication and behaviours (as provided by the 'Social Consequences' sub-advisory group of Exercise Ruaumoko; Smith [2009]).

Incorporating a wide range of expertise into an advisory group process is closely related to the concept of 'post-normal' science (Funtowicz & Ravetz [1991]; Krauss et al. [2012]) which is a 'new conception of the management of complex science-related issues' (Funtowicz & Ravetz [2003], p. 1) where 'facts are uncertain, values are in dispute, the stakes are high and decisions urgent' (Funtowicz & Ravetz [1991], p. 137), particularly when these uncertainties are of an epistemological or ethical kind. Post-normal science considers these elements of uncertainty, value loading, and a plurality of legitimate perspectives to be integral to science, and that by adopting this 'post-normal' approach there is a recognition that risks are interpreted and managed subjectively (depending on local values and norms as well as disciplinary frameworks). This approach presents a new problem-solving framework and acknowledges that a plurality of perspectives should be structured into the informed decision making processes during uncertain high risk environments (see WSS Fellows on RIA [2014]).

THE DEVELOPMENT OF VOLCANIC SCIENCE ADVISORY GROUPS IN NZ

One of the earliest exercises conducted to explore the response to a volcanic eruption in New Zealand was Exercise Nga Puia in 1992 (Martin [1992]). This exercise, based on a simulated eruption at the Okataina Volcanic Centre (Nairn [2010]), informed the response plans for the region, in particular the subsequent volcanic alert level processes. A few years later, these lessons were tested during the unrest and eruptions of Ruapehu volcano in 1995–1996, when over 42 organisations responded, with the Institute of Geological and Nuclear Science (now GNS Science) acting as the major science provider (Johnston et al. [2000]) (Figure 1).

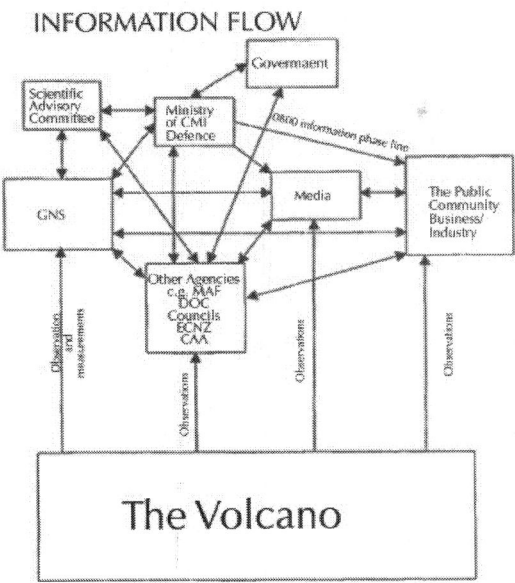

Figure 1: The information flow during the response to the 1995 Ruapehu Eruption showing information flow between key agencies (Paton et al. [1999]).

Analysis of the organizational response to these eruptions (Paton et al. [1998a]) identified the prominent role the limited formalised inter-organisational networking played in creating a coordinated response, particularly with regard to issues arising from ad-*hoc* interaction between science and response agencies. Response agencies became inappropriately over reliant on science agencies for management information. Paton et al. ([1998a]; [1999]) found that response agencies expected the geophysicists and volcanologists analysing volcanic activity to provide them with direct answers to all response management issues. For example, one co-author (DJ) reported how response agencies expected volcanologists to be able to answer questions about the effect of volcanic ash on sheep. Response agencies were unprepared for the need to be able to liaise with different expert sources (e.g. volcanologists, agricultural scientists, and veterinarians) to integrate input for multiple sources to make response management decisions. Similar problems emerged with regard to public health, environmental health, and utility issues. In addition to expecting certainty about volcanic hazard characteristics and future activity, response problems emanated from lack of attention in the response planning context given to understanding what sources of expertise would need to be consulted and how response management would need to integrate and interpret information while collaborating with diverse others. At the same time, the lack of networking meant that many agencies were unable to fully utilize scientific data as agencies found it was inconsistent with (unexpected) situational awareness and decision demands, discussed further in Section 3.

As a result, science advisory processes were redeveloped through the formation of a number of VSAGs (Smith [2009]; Jolly and Smith [2012]). During many natural hazard events, such science advisory bodies have been called upon to provide information and advice. The ability to source science advice through "one trusted source", such as the VSAG, has proved beneficial (Ministry of Civil Defence and Emergency Management, MCDEM [2008]) (Figure 2a). This approach also facilitates an integration of a wide range of expert opinions required to manage uncertainty during decision making (Lipshitz et al. [2001]) and can help combat issues arising from conflict between scientists (Barclay et al. [2008]).

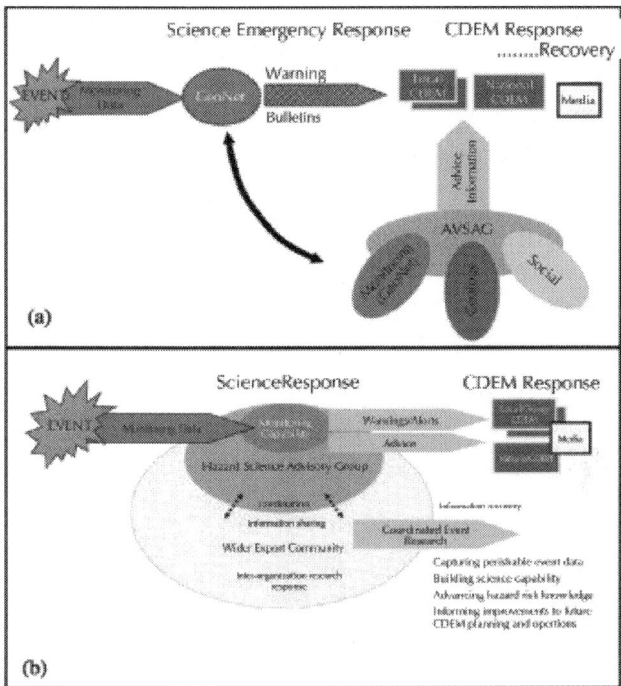

Figure 2: Information flow processes during a NZ Volcanic crisis. (a) The information flow during Exercise Ruaumoko in 2008, outlining the operating structure of the Auckland Volcanic Scientific Advisory Group and its relationship to GeoNet and CDEM (Smith [2009]). (b) The proposed model for a national hazard science advisory group, to enable integration of nationwide science capability, as proposed by Smith ([2009]) after Exercise Ruaumoko. Smith ([2009]) states that "this advisory group would be made up of appropriate subject experts from across universities, crown research institutes and other science organisations including consultancies, [and] ... could play both an operational role (during events) and a strategic role for planning science activities" (p. 76). The current advisory structure of the CPVAG, TVSAG, and CAG reflect a similar advisory process (Jolly and Smith [2012]).

These VSAGs represent advisory bodies that are on standby, and have plans to respond to a crisis or unrest period. The advice provided by VSAGs is vital for effective emergency management planning, intelligence gathering, and decision making and for the protection of life, infrastructure and welfare, and depending on local procedures, the VSAG may exclusively include scientists (volcanologists, meteorologists, etc.) or also include local and regional emergency managers and officials. In NZ, the earliest formalised VSAG was the Egmont Volcanic Advisory Group formed in the early 1990s alongside the deployment of the Taranaki Volcano Seismic Network (for a full history see Bayley [2004]), which meets once a year to review the monitoring data and other scientific research. By 2004, this group comprised representatives from Massey University, University of Auckland, GNS Science (NZ Crown Research Institute for "Earth, geoscience and isotope research and consultancy services", GNS Science Website [2014]), Department of Conservation and Taranaki Regional Council; and sat at the advisory group level of the Taranaki Civil Defence Emergency Management Coordinating Executive Group.

This group advised Taranaki Regional Council in the development of its first Volcanic Contingency Plan in 2000, which addressed the framework of Scientific Alert Levels, the principal emergency management activities for response, the expected hazards and the monitoring network. Nowadays the Egmont Volcanic Advisory Group has evolved into the Taranaki Seismic and Volcanic Advisory Group (TSVAG), encompassing additional representatives from Victoria University of Wellington, MCDEM, the Earthquake Commission, and local CDEM groups (TRC [2013]).

The experiences of the TSVAG have helped inform the process and formation of a number of other VSAGs throughout NZ, many of which have overlapping membership in terms of volcanologists and national level CDEM and response organisations, most of these contain representatives from local CDEM, response and lifeline organisations and scientists from across the NZ Universities, and Crown Research Institutes (CRIs) including GNS Science:

The Auckland Volcanic Scientific Advisory Group (AVSAG) was established by the Auckland Civil Defence and Emergency Management Group in 2007 as part of its Volcanic Contingency Plan and in preparation for the MCDEM led volcanic Exercise Ruaumoko, to provide advice to officials about the volcanic field residing under Auckland City (MCDEM [2008]; McDowell [2008]; Smith [2009]). This built on the pre-existing VSAG mechanisms established in the 2002 Contingency Plan (Beca Carter Hollings & Ferner [2002]) which included volcanologists (from GNS Science, Auckland Regional Council, and University of Auckland), meteorologists, and specialist medical advisors. In 2007, AVSAG was the first NZ VSAG to establish formal terms of reference that were signed by member organisations (Cronin [2008]), and this updated VSAG incorporated a wider range of representatives from Auckland, Waikato and Massey Universities, GNS Science, MetService, the Kestrel Group, as well as local and national CDEM representatives (Smith [2009]).

The Central Plateau Volcanic Advisory Group (CPVAG) was established in 2008 to "provide a forum for the collective planning and readiness activities for volcanic hazards in the Central Plateau" which includes the volcanoes Mt. Ruapehu, Mt. Ngauruhoe, and Mt. Tongariro (CPVAG [2009], p. 5). This group formed directly in response to the dam break lahar that occurred from the Ruapehu Crater Lake in 2007, and the recognition by key stakeholders that a combined expanded advisory group was needed for effective planning, preparedness, relationship building, and inter-agency coordination (ibid). CPVAG encompasses a Science Focus Group, a Planning Focus Group and a Communications Focus Group, all guided by a framework strategy and Contingency Plan, and who meet every six months to report back on work programmes, outcomes, and future plans. The processes set up by CPVAG were recently tested in 2012 during the Te Maari eruption and unrest period, and the evaluation of that response is currently ongoing as the eruption represents a critical opportunity to review effectiveness, and identify areas for improvement and capacity building.

The fourth VSAG in NZ is the Caldera Advisory Group (CAG, Waikato Regional Council [2014]; Potter et al. [2012]), which was formed in late 2010, with a focus on the eight caldera volcanoes in the Taupo Volcanic Zone. This group formed in response to the recognition of a gap in the advice provision available for the particular effects of caldera volcanoes, and an acknowledgment that these effects could last for significant time periods (years to decades) with a "profound impact on the social and economic environments" (Waikato Regional Council [2014]).

Similar science advisory group processes also exist for other hazards in NZ including the Tsunami Expert Panel, which activates in response to a local, regional, or distant source earthquake and tsunami warning. This advises officials of coastal regions at risk, expected tsunami arrival times and durations, and the expected maximum wave amplitudes at the coast, providing advice directly to the Ministry of Civil Defence Emergency Management response team (MCDEM [2010]).

More recently, during the 2010–2012 earthquake and aftershock sequence in the Canterbury region, the Natural Hazards Research Platform assumed the role of national coordinator of science advice when the government declared a State of National Emergency after the fatal M6.3 event in February 2011 (Canterbury Earthquakes Royal Commission – Te Komihana Rūwhenua o Waitaha[2012]). This government funded multi-party research management platform was established in 2009 to provide secure, long-term funding for natural hazard research, to encourage stakeholder involvement in research, and to promote collaborative research (Natural Hazards Research Platform [2009]). In future natural hazards events, and based upon the Canterbury experiences, the various scientific advisory group sections of the wider volcanic advisory groups described above would likely fall under the coordination of the Natural Hazards Platform, or a future equivalent, as they fulfil their science advisory role; pending reviews of recent volcanic eruptions and earthquake events of the last 4 years, and changes currently under consideration for new national funding procedures.

AVSAG was the first VSAG to be comprehensively tested in a simulation. This was done through Exercise Ruaumoko, which was a MCDEM led exercise to test an all of nation response to a volcanic eruption in the Auckland volcanic field (MCDEM [2008]) and was the first test of AVSAG. For this, AVSAG was co-ordinated through a tri-partite sub-group system (Volcanology, Volcano Monitoring, Social Consequences), which reported upwards to a smaller core VSAG that liaised directly with MCDEM and Auckland CDEM via on-site liaison officers in the Emergency Operation Centres (EOCs) at each location (see reviews in: Smith [2009]; Doyle & Johnston [2011]).

Reviews of this exercise (MCDEM [2008]; McDowell [2008]; Cronin [2008]) identified that the AVSAG process facilitated the provision of valuable advice in a clear, timely manner. As advocated for by the International Association for Volcanology and Chemistry of the Earth's Interior ([1999]), the AVSAG provided a facility for the scientists (from all contributing disciplines) to "use a single voice", share information to reduce confusion, and to encourage efficient teamwork amongst scientists and public officials, while also encouraging integration of diverse scientific expertise and minimising communication delays. However, during the most active periods of the response towards the end of the exercise, the existence of two distinctly separate scientific sub-groups composed of the predominately university-based 'volcanology' group and the 'monitoring' group of GNS Science based scientists became unrealistic, and as stated by McDowell ([2008]), p. 22, the priority instead should be to have "the rapid assessment and decision-making in relation to technical data" rather than maintaining and communicating between these two separate groups. While the main advantage of this AVSAG approach was the wide range of scientific experts and competency, during the most active period the due process needed to maintain this inclusivity actually slowed down advice provision (MCDEM [2008]).

A clear advantage during the exercise was the presence of a science advisor in both the National Crisis Management Centre (NCMC) and the Auckland CDEM Group EOC, providing a vital link between the VSAG assessment and the emergency management decision-making. However, during the rapid escalation of unrest and the critical moments of the crisis, there was a potential for disconnect to occur between this local and national advice provision to the AGEOC and the NCMC respectively, and this resulted in a divergence of the science advice (which was informing evacuation planning) at Local and National levels (Cronin [2008], discussed further in Section 5.2).

A potential disconnect could also occur not only between the science advice at the local and national CDEM level, but also between the local and national science research response, capability, and processes in future events. Smith ([2009]) suggests that to address these limitations and coordinate the scientific advice beyond the limited knowledge pool and resources in a locally impacted area, a nationwide Volcanic Science Advisory Panel (NZVAP) should have (see Figure 2b):

"at its core national hazard monitoring capability and processes (e.g. GeoNet), with involvement of additional capability from universities and other science organisations based on thresholds of response. The intent is that GeoNet (both the technology and the science expertise of GNS Science) be the hub of any science response for earthquake, volcano, tsunami or landslide events" (p. 77).

GeoNet is a collaboration between the Earthquake Commission (EQC - NZ's insurance provider for natural disasters; EQC Website [2014]) and GNS Science and is the "official source of geological hazard information for New Zealand". Established in 2001 it monitors geohazards (in particular earthquakes, tsunami, volcanoes and landslides) via an extensive monitoring system and archival data centre, and provides public and official information including earthquake reports and Volcanic Alert Bulletins (GeoNet website [2014]). The NZVAP approach outlined by Smith ([2009]), encompassing GeoNet at its core, still supports the existence of regions having existing scientific or planning advisory groups with a volcanic and/or earthquake focus, but it also addresses the need for mobilisation of NZ-wide science capability, while remaining responsive to local CDEM needs (see also Section 5.2; Jolly and Smith [2012]).

Developing VSAGs prior to an event can prospectively enhance the crisis response capability of scientists and the full multi-agency response alike, through the development of terms and protocols for response, information sharing and inter-agency management, and situational awareness. A prospective approach facilitates future collaboration (e.g., enhanced mental models, shared situational awareness, enhanced multi-team performance) and more effective communications between scientists and responding agencies.

FACTORS THAT INFLUENCE RESPONSE EFFECTIVENESS

Even with a pre-existing VSAG, the development of effective multi-agency response needs to accommodate issues arising from, for example, differences in organisational culture, jurisdictional expectations, and differences in the economic and political pressures on participating agencies. These represent demands on the goal of prioritising tasks and information needs over and above those emanating from the complex, evolving volcanic hazard. The corresponding threats to trust, leadership or team ability, conflicts over responsibility or priorities, reputation management, and need to function under high psychological and environmental stressors (fatigue, tunnel vision, family commitments and over work) conspire to impair performance of the individual and team (Boin and't Hart [2001]; Handmer [2008]; Quarantelli [1997]; Sinclair et al. [2012b]; Paton[1996]). In addition, conflict may arise as individuals swap hierarchal/seniority position as they move from their day-to-day role to their response role.

To manage these issues, it is essential to build future response capability via the development of good team and inter team mental models. This facilitates situational awareness and enhanced decision making capability for personnel within the VSAG and key responding agencies. This should be supported by training (Sections 4 and 5) and resource planning (e.g., accommodating the need to coordinate multiple shifts throughout a response, manage fatigue, allow individuals to attend to personal and family demands) that ensures that staff can return to their role with a fresh perspective. Limiting time on shift has pragmatic benefits in reducing response risks from personnel adapting to small incremental change and losing situational awareness (e.g. Tickell[1990]). In contrast, a fresh responder/scientist would recognise a significant change demanding an action. This tendency, colloquially referred to as the 'boiling frog syndrome'[a], was particularly noticed in the volcanologist's experience when managing the 1991 eruption at Mt. Pinatubo (Whittlesey and Buckner [1993]), and highlights the need for regular shift rotation of scientists and other responders in crisis response.

[a]"The "boiled frog" scenario is referred to often in disaster and business communities to describe "creeping" disasters and crises, it describes a frog that when placed in a pot of cold water which is being gradually heated, will fail to recognize the increasing danger and thus get boiled alive. However, if this proverbial frog is put directly into hot water, it will recognize the dangerous situation and jump straight out of the pot. This metaphor thus describes a number of problems that can arise with creeping disasters: failing to recognize the accumulation of many small changes which can amount to a major crisis; normalization bias; and personnel fatigue and performance issues for long duration events". (Doyle et al. [2014b], p. 86).

It is also important to consider rotating the role of the lead science individual (or agency) within a VSAG during a response, particularly for long duration crises. In a NZ context for example, many of the responding scientists will be balancing their 'response role' (to provide advice to responding agencies and understand the phenomena occurring) with their 'day-to-day role' (including funded research, consultancy contracts and lecturing). The 'day-to-day role' may need to take priority at times, and thus by rotating the role of 'lead' individual (or agency), this will allow time and space to respond to these other competing demands, while also managing issues around fatigue and tunnel vision as the various agencies and individuals 'share the load'. Further, by rotating these roles, individuals can develop a greater understanding of each other's responsibilities, roles, pressures, and demands, helping to build a better shared mental model of the team response (see Sections 3.3 and 4.1). For any role swap or shift change to work effectively an effective change-over procedure (e.g., role shadowing for a specific time period) is needed to transfer situational awareness and overall response performance. This is essential to maintain decision making effectiveness in evolving crises.

Decision Making

Pivotal to effective volcanic crisis management are the decision making, situational awareness, mental models (which is an individual's representation or visualisation of a real system, including concepts, relationships, and their role within that system) and trust processes that underpin effective response. We commence discussion with an overview of the individual and group decision making processes occurring in volcanic management and response.

Analytical decision-making is defined by working through a process: identifying a problem; generating options to solve the problem; evaluating these; and implementing the preferred option (Flin [1996]; Saaty [2008]). This form of decision making requires time to allow this process to occur. It is the default approach adopted by scientists (and managers) due to their training. However, in emergency response a range of other decision making styles: analytical; naturalistic; procedural based; creative; and distributive (e.g. Crichton & Flin [2002]); are required and need to be matched to the situation and conditions encountered by decision-makers.

The slower, more considerate, analytical decision making processes lie at one end of a continuum of decisions styles. At the "faster" end of the decision making continuum lies *naturalistic* decision making (NDM; Martin et al. [1997]). This relies on experience garnered through real world crises as well as simulations and exercises (Crichton and Flin [2002]; Klein [2008]). It is commonly adopted in high risk and low time contexts and *naturalistic* settings which involve: ill-structured problems; uncertain, dynamic environments; shifting, ill-defined, or competing goals; action/feedback loops; time stress; high stakes; multiple players; and the effects or pressures of organizational goals and norms (Orasanu & Connolly [1993], as cited in Zsambok [1997], p. 5). For critical incident management, research has identified four key NDM processes (Crego and Spinks [1997]; Crichton and Flin [2002]; Pascual and Henderson [1997]): 1) recognition-primed and intuition led action; 2) a course of action based upon written or memorised procedures; 3) analytical comparison of different options for courses of action; and 4) creative designing of a novel course of action; ordered by increasing resource commitment.

In a crisis, uncertainty, environmental change, risk and time pressures are amplified, making decision making (whether by scientists or responders) in this dynamic context one that is dependent on 'task conditions' (Martin et al. [1997]), and thus throughout one incident different processes may be adopted. For example, during Exercise Ruaumoko (Sections 2.1 and 5.2), the on-site GeoNet duty officers inside both the Auckland Group EOC and the NCMC were often asked to provide answers to questions from key decision makers who required an immediate response due to response management demands. This "task condition" (short time) would have favoured the more *naturalistic* decision making, where the scientists would have relied upon their experience to assess the situation (both the question and the available science) to make an intuitive or recognition-primed decision (Paton et al. [1999], Klein [1998]). However, earlier in the exercise, during the warning period preceding the volcanic 'crisis' the science advice could be carefully evaluated and compared, and relatively lower time pressures afforded scientists and decision makers the opportunity to adopt an *analytical* decision making approach. However, as the situation moved from this early warning phase into the crisis of impending eruption, or for a situation where scientists are responding rapidly after an eruption (e.g. the "Blue Sky" eruption of Ruapehu in September 2007), the more intuitive naturalistic style would again be adopted. As stated by Paton et al. [1999], "attention must [thus] be directed to understanding [this] naturalistic decision-making of experts, and how it can be modelled in simulations to develop this contingent management capability" (p. 44).

Situational Awareness

Pivotal to effective decision-making process is a capacity to: 1) evaluate and define a problem and task characteristic via situation assessment (Endsley [1997]; Martin et al. [1997]); and 2) select a decision-making strategy from the four options described above (Crichton and Flin [2002]). The former process is intrinsically dependent upon the situational awareness (SA) of the individual and the team (Cannon-Bowers and Bell [1997]; Crichton and Flin [2002]).

Situational Awareness comprises three levels (Endsley [1997], p. 270–271): 1) Perception – understanding the importance of information and cues in the environment; 2) Comprehension – combine, interpret, store, and retain information and be able to use it; and 3) Projection – prediction of future situations from existing and previous situations. Initial and ongoing SA is critical to decision making (Sarna [2002]). Thus a decision-maker may make the correct decision based upon their perception of the situation, but if their situation assessment is incorrect then this may negatively influence their decision (Crichton and Flin [2002]). When responding to a volcanic eruption, developing and maintaining this SA is important for both volcanologists (in their assessment of available data and future projections, information from other agencies, demands upon their advice) and emergency managers and decision makers (in their assessment and understanding of information - including science advice, resources, demands and future requirements and needs).

In reviews of the inter-organisational response of the 1995–1996 Ruapehu eruptions in NZ (Paton et al. [1998a], [1998b], [1999]), several issues affected the SA of both scientists and emergency managers. In particular, "inter-organisational networking" was weak, with none of the responding agencies (e.g. fire, police, civil defence, social services, media, etc.) having an established or formalised inter-organisation network in place with GNS Science before the event, even though GNS Science acted as an information provider for 63% of the responding agencies. This resulted in organisations interacting on an "ad hoc" basis (Paton et al. [1998a], p. 7) contributing to co-ordination and communication problems and preventing their using crucial information to build and maintain SA, and thus impacting both emergency management and volcanological decision-making processes and outcomes. This was compounded by issues such as the "lack of clear responsibility for co-ordination" across responding agencies (reported by 45% of participating agencies), "inadequate communication with other agencies" (37%), and "inadequate co-ordination of response" (32%). These are all indicators of "team breakdown", inadequately defined and co-ordinated roles, and poor communication (Paton et al. [1998a], [1999]). The development of effective inter-organisational crisis communication requires (Paton et al. [1998a]) "information needs [that] are anticipated and defined, that networks with information providers and recipients are organised, and [that] crisis communication systems capable of providing, accessing, collating, interpreting and disseminating information are established. … [and] shared terminology and systems" (p. 8).

The development of VSAGs and the associated Terms of Reference within NZ over the last two decades will have helped to build shared mental models of the response environment across organisations. VSAG development has included the identification of protocols for communication and networking with emergency management and key response organisations, and specifying relationship building activities to be undertaken within these groups. These activities have improved communication and information flow during a volcanic crisis and thus the shared situational awareness in that crisis. Comparison of Figures 1 and 2 illustrates the change and improvement in information flow processes between Ruapehu in 95–96 (Figure 1) and Ruaumoko in 2008 (Figure 2).

The rarity of large scale eruptions makes it important to capitalize on the learning opportunities events, exercises, and reviews provide for developing situational awareness and for facilitating the ability of scientific advisors to develop shared mental models of their and others role in response management. This feeds into training needs analysis and the development of the situational awareness competencies and decision support systems required to sustain effective situational awareness in complex, rapidly evolving and dynamic volcanic crises. It can also inform the development of the shared situational awareness required if all team members make their respective contributions to a shared task or goal. That is, to develop shared mental models of presenting problems and response options, particularly when decision inputs come from different professions and/or from participants who are spread over a large geographical area. Facilitating the latter introduces a need for distributed decision making, discussed next.

Shared Mental Models

The scale and complexity of volcanic crisis response results in decision-making involving people who differ in their profession, expertise, functions, roles and geographical location. This more integrative decision style is called distributed decision making (Rogalski & Samurcay [1993], as cited in Paton & Jackson [2002]). As discussed in Section 3.1, an individual's mental model impacts individual decision making, as it is their representation of the wider system and processes, including inter-agency relationships, needs and demands, and an individual's role within a crisis. Thus, for effective *distributed decision making*, individuals require a good *shared mental model* of the response environment in time and space, which incorporates how their expertise contributes to different parts of the same plan, and their understanding of each other's knowledge, skills, roles, anticipated behaviour or needs (Flin [1996]; Marks et al. [2002]; Paton and Jackson [2002]; Schaafstal et al. [2001]). By building a shared mental model, team members can develop an accurate expectation of the performance of their team members and themselves, leading to effective coordination without overt strategizing (Blickensderfer et al. [1998]; Cannon-Bowers et al. [1998]; Lipshitz et al. [2001]; Salas et al. [1994]).

The collection of GNS Science information by response agencies in an *ad-hoc* basis (Paton et al.[1998a] & Paton et al. [1999]) during the 1995–1996 Ruapehu eruptions describes a process where information was being provided by explicit requests only. In such cases, the information provided often also needed to be adapted and translated to meet decision needs. However, research in the decision making community has identified that effective teams move from the sharing of information by *explicit request* towards an approach that adopts *implicit supply*, where members provide not only good information, but unprompted information that is tailored in terms of content and format due to their understanding of the needs of the recipient (Lipshitz et al.[2001]; Kowalski-Trakofler et al. [2003]; Paton and Flin [1999]). *Implicit* communication also facilitates the maintenance of situational awareness during periods of dynamic information, as it allows decision makers to focus on task management. For this kind of team functioning to be successful in complex, time pressured situations, (Wilson et al. [2007], as cited in Owen et al.[2013], p. 5) identified that it required the following characteristics:

Effective communication consisting of accurate and timely information exchange, correct phraseology and closed-loop communication techniques;

Coordinated behaviour based on shared knowledge, performance monitoring, back-up and adaptability; and

A co-operative team orientation, efficacy, trust and cohesion.

However, as highlighted by Owen et al. ([2013]), p. 6 for the multi-team, multi-organisational, coordination characteristic of large-scale complex emergency management events, the challenge is not just to build an effective team, but "to understand how a team coordinates *within* teams" and how understanding may be "shared *between* teams". When we consider a multi-agency VSAG, the responding scientists can be considered as but one team within the complex multi-team response. Owen et al. ([2013]) identify four distinct stages for effective and adaptive team functioning for an inter-team inter-organisational response, including: 1) situation assessment, 2) plan formulation, 3) plan execution, and 4) team learning. These, and the indicators typically used to identify whether these activities are occurring, are depicted in Table 1.

Table 1: Framework for inter-team inter-organisational coordination (Owen et al.[2013])

Phase	Within teams	Between teams	Anchor points
Situation assessment	Information gathering, individuals scan the environment to identify cues	Boundary spanning	Within teams: incident briefings; handovers
	Individual and Team Situation Awareness	Distributed Situation Awareness Social networks	Between teams: Emergency Management Team (EMT) briefings; situation reports; emergency services liaison officers
	Sense-making	Organisational culture	Information flows through texting, emails, data retrieval

Plan formulation	Meaning making	Shared beliefs	Regional and state level team membership
	Setting goals, clarifying roles, prioritising tasks	Centralised-decentralised decision making authority	Decision-structures analysis
	Psychological safety	Distributed cognition	Command and Control (C2) teleconferences; EMT meetings
	Team trust and cohesion	Social networks	Regional and state level team membership
Plan execution	Communication	Relational coordination	Observations of teamwork
	Explicit and implicit coordination	Cultural-historical activity theory	Temporal and cultural-structural boundary points
	Cross-checking/ monitoring/ backup behaviour		
	Leadership	Boundary spanning	C2 teleconferences; EMT meetings
Team learning	Psychological safety	Analysis of organisational tensions and contradictions	Within teams: immediate debriefs
	Opportunities for reflection and perspective-taking	Organisational learning (post response)	Between teams: multi-agency after action reviews; development of knowledge networks.

Doyle *et al.*

Doyle *et al. Journal of Applied Volcanology* 2015 4:1, doi:10.1186/s13617-014-0020-8

A significant challenge here derives from a need to coordinate the inputs of different agencies and experts to assist the holistic management of complex hazard consequences. For example, public health specialists possess expertise concerning the specific effects of ash and gas on health. However, to mount an effective response, their input, as members of an 'emergent' team, must be integrated with input from, for example, volcanologists, emergency managers, social welfare, and transport agencies, to facilitate understanding of the 'whole' problem, prioritise issues, and to identify where is safe for people to be evacuated to, etc. This example illustrates how, for example, public health specialists and volcanologists need to bring their professional team mental model to bear on identifying their specific contribution, but also develop a superordinate (overarching) mental model that integrates all areas of expertise. A significant challenge arises because scientists, and indeed all stakeholders, need to switch between a) being autonomous actors, and b) being multi-disciplinary team members, depending on the task being undertaken (Janssen et al. [2010]).

Fundamentally, in a volcanic crisis response setting, the science advisors' and VSAGs' role is to provide information to facilitate the response agencies understanding of the hazard issues, priorities, the wider context, impacts, and potential future outcomes, and thus build their situational awareness to aid their decision-making process. However, as stated by Doyle & Johnston ([2011]), it is not just a case of providing the emergency managers with all available science information, but about understanding their needs to meet their information requirements. Simply providing as much advice as possible may actually hinder the decision process, due to cognitive overload and an overuse of these available resources (Crichton and Flin [2002]; Omodei et al. [2005]; Quarantelli [1997]). To contribute in this way, scientific advisors need to develop a shared mental model with their emergency manager counterparts both prior to an event (to develop effective plans) and during the crisis itself. This shared mental model encompasses the overlapping elements of each team member's SA and represents the inter-team co-ordination (Endsley[1994]), allowing an effective coordination amongst team members without the need for extensive overt strategizing (Salas et al. [1994]; see review in Doyle & Johnston [2011]).

For science response it is thus vital that scientists develop this shared mental model within the VSAG, and in the wider multi-agency response, to facilitate their ability to implicitly provide the science information required by the main decision makers at critical periods (see also Doyle & Johnston [2011]). However, when dealing with the uncertainty implicit in volcanic crises the effectiveness of this information sharing relationship is influenced by the degree of trust that exists, or that needs to be developed in situ, between key players. This is particularly important given the rarity of opportunities for functional organizational interaction before a crisis occurs.

Trust

Trust plays a pivotal role in developing sustainable, functional relationships when members of diverse organizations need to collaborate to access, share and use information for decision making in response environments characterized by uncertainty (Siegrist and Cvetkovich [2000]). Without trust, teams focus on task demands, not teamwork, reducing their effectiveness in tackling emergent crisis response needs (Pollock et al. [2003]).

Inter-agency trust develops through collaboration. Since volcanologists and emergency managers rarely work together under normal circumstances, trust among agencies may be lacking (Dirks and Ferrin [2001]). Since representatives of scientific and EM agencies typically meet and interact for the first time during a crisis, agency representatives are denied the luxury of building trusting relationships over time. Trust must be developed via other mechanisms. One approach capitalizes on the concept of swift trust (e.g. Meyerson et al. [1996]).

Swift trust can be developed in temporary (EOC) organizations if certain conditions are met. Meyerson et al. ([1996]) argue that, firstly, swift trust is less about the interpersonal relationship factors that underpin traditional forms of trust (built up over a prolonged period of time), and more about encouraging a focus on goal achievement by facilitating the ability of participants to understand their respective contributions to a superordinate (overarching) team managing complex evolving eruption consequences. Secondly, swift trust is more likely to arise when drawing upon a small pool of representatives who have an increased chance of future interaction, with this condition creating a social setting that can foster quicker trust building between parties. Finally, swift trust avoids personal disclosure in favour of a reliance and focus on key tasks that relate to the features of the setting (i.e., the need to integrate diverse organizational and professional perspectives to tackle specific response issues as a member of a superordinate team). If all members of the superordinate EOC organization, incorporating the VSAG, adopt these roles they are more likely to be able to develop trusting relationships that facilitate effective information exchange and utilization in high risk, evolving crisis events.

Evidence of swift trust first emerged from Goodman and Goodman's ([1976]) observation that some temporary groups did not have a history of trust, but developed "swift trust" through task related interaction. Evidence for the effective role that swift trust can play in multi-agency and distributed management systems comes from research into global virtual teams that exemplify temporary organizations; and the requirement for collaborative team management further supports its utility (Coppola et al. [2004]; Crisp & Jarvenpaa [2013]; Robert et al. [2009]; White et al. [2008]). The concept of swift trust has only recently been tested in multi-agency natural hazard crisis response contexts (Curnin et al. [2015]). This, as well as evidence for its effectiveness in military contexts which involve the collaborative response to emergent, low-time/high-risk demands over time (Ben-Shalom et al. [2005]; Hyllengren et al. [2011]; Lester & Vogelgesang[2012]) suggests it should be included in future volcanic crisis response protocols. Swift trust research in military contexts also highlighted the importance of selecting team members with sufficient status to be 'heard' in a multi-agency team context (Curnin et al. [2015]).

ACTIVITIES TO IMPROVE FUTURE RESPONSE CAPABILITY

In the above discussion, we outlined that effective individual and team response to a crisis, such as a volcanic eruption or unrest period, is characterised by good situational awareness, strong inter-organisational networks, effective shared mental models, and high trust between responding organisations and individuals. To achieve this, and develop a common understanding of each other's roles, dependencies, and information needs, and the over-all response environment, it is important to undertake multi-organisational and multi-disciplinary planning activities, and collaborative exercises and simulations with all team members and advisors, to help in the development of similar mental models of the task (see review in Paton & Jackson [2002]; Doyle & Johnston [2011]). Such a comprehensive suite of training and relationship building activities prior to an event, and detailed analysis of event and exercise response, can facilitate future response capability and identify areas for improvement. This is particularly important given the rarity of volcanic and other hazard events, and thus a lack of opportunity for real world experience.

According to Kozlowski ([1998]), p. 120–122, team training should be considered as a sequence or series of developmental experiences that are carried out across a series of different environments, to build "knowledge and skills in an appropriate sequence across skill levels, content and target levels". Ideally this training and exercising needs to develop both individual and team situational awareness (SA) and explore how and when each is appropriate for response, within evolving, dynamic response environments. Team SA can be developed in post-event and post-exercise reviews that include identifying inter-agency relationship issues as opportunities for development (and not as problems). Through the analysis of past events, lessons for successful communication, advice provisions and distributed decision-making can also be learnt. However, these training activities need not necessarily develop shared mental models and the capacity for true levels of collaborative management etc. Agencies can also use them to update, write, and prepare plans, identify potential or existing issues with the response, logistical, and communication plans; while also testing such processes, systems, and communications. Adopting a suite of training activities increases opportunities for developing an understanding of the technical issues involved and the multi-agency context in which they occur (Borodzicz & van Haperen [2002]).

Several training methods have been identified that can enhance naturalistic decision-making (e.g. Cannon-Bowers & Bell [1997]), enhance decision skills (e.g. Pliske et al. [2001]), train effective teams (e.g., Salas et al. [1997b]), and develop effective critical incident and team based simulations (e.g. Crego & Spinks [1997]; see review in Flin [1996], chap. 6), all of which are relevant for volcanologists and VSAGs. These include cross training, positional rotation, scenario planning, collaborative exercises and simulations, shared exercise writing tasks, and 'train the trainer' type tasks; in addition to workshops, seminars, and specific knowledge sharing activities.

We briefly discuss below two methods in particular: cross-training and scenario planning, as we feel that they are particularly suitable for volcanic response environments. Exercises, and the application of them to science response, are discussed in detail in Section 5, including an evaluation of the lessons learnt from Exercise Ruaumoko in the context of the key competencies discussed in Section 3. It is important to highlight that for all these, it is not just knowledge and skill development that is addressed through these activities; they also address "how the disaster context influences performance and well-being" (Paton et al. [2000], p. 176). In addition, each of the training activities can be carried out at the many levels of a response, for teams within an agency, for the entire organisation, across multiple organisations, and for the full multi-organisation response.

Cross Training

Cross training enhances the awareness and knowledge that each team member has of their fellow team members' tasks, duties and responsibilities and facilitates the holistic (shared mental model) understanding of team functioning and the respective, interdependent role of a given agency within the team (Marks et al. [2002]; Schaafstal et al. [2001]; Volpe et al. [1996]). This is termed their *interpositional knowledge* (IPK). Volpe et al. ([1996]) reason that IPK allows team members to "anticipate the task needs of fellow team members", leading to more effective team performance and enhanced coordination with a minimal communication requirement, important when task loads are high and individuals are too busy attending to these to be able to make explicit information requests. In the absence of IPK, there exists *interpositional uncertainty* which can hamper team performance (Blickensderfer et al. [1998]).

Cross training is "an important determinant of effective teamwork process, communication, and performance" (Volpe et al. [1996], p. 12). Teammates who develop interpositional knowledge through cross-training: 1) interacted more effectively with each other, 2) used more efficient communication strategies, and 3) volunteered information more often (Blickensderfer et al.[1998]). It facilitated these outcomes by "encouraging members to understand the activities of those around them" (Blickensderfer et al. [1998]), to better anticipate their needs and assist those in need of help (see Table 2 and Schaafstal et al. [2001];

Marks et al. [2002]). Furthermore, cross-training can foster a sense of a shared "common bond" (Greenbaum [1979], as cited in Blickensderfer et al. [1998]) amongst team members and support the establishment of morale, cohesion and confidence.

Table 2: Lessons learned from studies on cross-training teams (Blickensderfer et al.[1998])

1	Cross-trained teams are better able to anticipate each other's needs (Volpe et al. [1996])
2	Cross- training fosters inter-positional knowledge (Cannon-Bowers et al. [1998])
3	Cross-training should be used in combination with team process training to provide maximum benefit
4	Length of cross-training intervention is not necessarily related to value of the intervention
5	Cross-training interventions should be designed on the basis of the interdependency requirements of the task, i.e. teams with high interdependencies be given positional rotation, whereas teams with few interdependency requirements may need only basic knowledge of team structure, through positional clarification

6	A number of guidelines regarding the training objectives and content can be based on the cross-training research. From these, cross-training should do the following (adapted from Salas et al. [1997a], Blickensderfer et al. [1998]):
	i. Provide team members with an understanding of how other team members operate, why they operate as they do, and the manner in which they are dependent on teammates for information and input;
	ii. Provide team members with exposure to the roles, responsibilities, tasks, information needs, and contingencies of their teammates' tasks;
	iii. Provide team members with practice on the roles and tasks of teammates, highlighting the interdependencies of the positions as requested; and
	iv. Provide feedback during cross-training exercises that allows teams members to formulate accurate explanations for their teammates' behaviour and reasonable expectations for their teammates' resource needs.
	v. To determine the specific content of cross-training, a team task analysis should be conducted. This will help to identify interdependencies in the task and to identify what inter-positional knowledge is necessary to help teammates coordinate

Doyle *et al.*

Doyle *et al. Journal of Applied Volcanology* 2015 4:1, doi:10.1186/s13617-014-0020-8

Cross-training encompasses three methods that become progressively more detailed and involved, and thus more effective for improving shared mental models, understanding of complementary roles, and enhancing collaboration (Blickensderfer et al. [1998]; Marks et al. [2002]). These are: 1) *positional clarification*, a form of awareness training (e.g., by discussion, lecture, demonstration or dissemination of information) where specific information is provided about other roles and responsibilities in the team (e.g., working together in an EOC, or in a science response team, for example); 2) *positional modelling*, a training procedure in which the duties of each team member are discussed and observed via behaviour observation (e.g., offering potential EOC participants some actual practice in the other positions: a volcanologist could be given the opportunity to play the role of, for example an intelligence or logistics officer in an EOC context); and 3) *positional rotation*, which involves training within the exercise context where all team members spend significant periods of time performing other team members' jobs and roles, providing a working knowledge of each member's specific tasks and how those tasks interact and to gain different perspectives of the overall situation. The chosen cross training method amongst these three approaches should correspond to the realistic level of interdependence of the team (Ford and Schmidt [2000]).

Scenario Planning

Scenario planning is a technique that creates multiple scenarios of "different futures" in ways that accommodate the perspectives of multiple agencies (i.e., to develop response scenarios that more accurately reconcile the needs, goal and expectations of diverse agencies; Bloom and Menefee[2014]; Moats et al. [2008]; Paton [2014]). At a fundamental level, scenario planning allows an "organisation to examine several options or risks that might have been overlooked in a plan that was constructed around a single environment. The process forces managers to think about the unthinkable and to even plan for it" (Bloom and Menefee [2014]). Through this process, the goal is to outline possible futures that are "credible and yet uncertain" (Keough and Shanahan [2008]). Various alternative steps to the scenario planning process are outlined in Table 3.

Table 3: The generic scenario planning model of Keough and Shanahan ([2008]), and an example of a scenario building model for step 3 given by Schwartz' 8-step approach (Schwartz[1996]; as cited in Keough and Shanahan[2008])

Generic scenario planning model		Schwartz' 8-step scenario building model	
1	Engage in scenario planning	1	Identify focal issue or decision
2	Compose the team	2	Identify key factors in the local environment which influence the decision
3	Scenario Building	3	Identify driving forces that influence key factors in the local environment
4	Decision process	4	Rank by importance and uncertainty
5	Increased performance	5	Select scenario logics
		6	Flesh out scenarios
		7	Consider implications
		8	Selection of leading indicators and signposts

Alternatives to this scenario building model, with additional steps: a) research, b) identifying major stakeholders, and c) communication, can also be found in Wilson and Ralston ([2006]) and Moats et al. ([2008]). The 18 step approach of Wilson & Ralston provides a clear road map through four phases of scenario planning, including a) getting started, b) laying the environmental analysis foundation, c) creating the scenarios, and d) moving from scenarios to a decision. Through these various approaches, the current mental models of participants and their assumptions can be identified and improved upon.

Doyle et al.

Doyle et al. Journal of Applied Volcanology 2015 4, doi:10.1186/s13617-014-0020-8

Scenario planning enables the integration and awareness of the various social, political, economic, cultural, and other environmental forces that underpin the histories and expectations different agencies bring to the response management environment. It also provides an opportunity to 'rehearse the future', promoting adaption versus reaction, and providing a safe space through which various points of view and new or unique ideas from within the team can be shared without the fear of being prejudged or automatically dismissed (Bloom and Menefee [2014]). Through this process, the volcanic crisis management team can also enhance their understanding of the wider response process, the key issue and decision thresholds, and trigger points; thus facilitating their collective ability to integrate their various perspectives and develop a bigger picture of the response than would arise if this was based on an individual role or stakeholder working separately. This can enhance shared situational awareness of the issues and process amongst the participating team. Scenario planning can be conducted at the sub-agency level (e.g. a monitoring team at a volcano observatory planning the response and deployment of personnel and instrumentation to facilitate effective monitoring during outcomes of an unrest scenario), at the agency level (e.g. a government science agency and university identifying how they could share resources and support each other during a response) and the multi-agency level (e.g. the VSAG considering potential eruption and response scenarios with the local CDEM group). In particular, for scientists, it provides an opportunity to work alongside emergency managers to identify science information needs and impacts of that information on decision outcomes, thresholds and trigger points.

EXPLORING EXERCISES IN DETAIL

In the field of emergency management, exercises to test response and procedures, and to train personnel commonly fall into one of the five categories listed in Table 4, from the small scale in-house orientation exercise to a full-scale multi-agency exercise conducted at the local, national and international level. These different types of exercises may then each be conducted at the different levels of the CDEM structure, for example in NZ they are (MCDEM [2009a]):

Tier 1: Local exercise (individual organisation)

Tier 2: Group exercise (within CDEM group)

Tier 3: Inter-Group exercise (across CDEM Groups, may include MCDEM)

Tier 4: National exercise (New Zealand or part thereof, including central government).

Table 4: The five types of emergency management exercises, as described by MCDEM ([2009a])

What	Details	Example
Orientation exercise	A 'walk through'. It puts people in a place where they would work during an event, or uses them as participants in a demonstration of an activity. This type of exercise is used to familiarise the players with the activity	Setting up a mock welfare centre, and walking staff through how it is organised.
Drill exercise	Players physically handle specific equipment or perform a specific procedure. The exercise usually has a time frame element and is used to test the procedures.	Activating an emergency operations centre or using alternative communications (such as radios).

Tabletop or Discussion exercise	Participants are presented with a problem that they are required to discuss and formulate the appropriate response or solution. Can be:	Participants discuss their response to a tsunami threat to a particular area, where the only injects are Tsunami Bulletins, Watches or Warnings from the Pacific Tsunami Warning Centre in Hawaii, describing the nature of the threat.
	• Facilitated: where the exercise controller actually facilitates the discussion through a series of questions in a stress free, relaxed environment, designed to identify gaps or problems in procedures or resources	
	• Inject driven: where the personnel are provided a scenario and prewritten exercise injects, to practice problem solving and co-ordination of services – with or without time pressures	
	In both of these, there is no actual deployment or use of equipment or resources.	

Functional exercise	Can also be called an operational or tactical exercise, it takes place in an operational environment and requires participants to actually perform the functions of their roles. A normally complex response activity is simulated, which lacks only the people "on the ground" to create a full-scale exercise.	A multi-agency response to extensive flooding, where evacuation of a settlement is required. Messages and injects are provided by exercise control and are handled by the participants in the way described in appropriate plans and procedures. Outcomes are generated that would be expected in a real situation.
	• Participants interact within a simulated environment through an exercise control group who provide prewritten injects and respond to questions and tasks developing out of the exercise.	
	• Functional exercises normally involve multi-agency participation (real or simulated) and it can focus on one or many geographical areas.	
	• This type of exercise is used to practice multiple emergency functions e.g. direction and control, resource management and communications.	

Full-scale exercise	Sometimes also called a 'practical' or 'field' exercise. These include the movement or deployment of people and resources to include physical response 'on the ground' to a simulated situation.	Deployment of a small team to a simulated car crash or industrial rescue by a single agency, using real rescue equipment.
	• They can be 'ground' focused only or may include the higher level response structures, and they can be simple (single agency) or complex (multi agency).	Or, coordinated multi-agency response to a tsunami warning involving actual evacuations and actors portraying the public.
	• These exercises are typically used to test all aspects of a component of emergency management.	
	• Can be simple single agency, or complex multi-agency.	

Similar definitions are provided by the US Department of Homeland Security, with further types including Seminars, Games and Drills (HSEEP [2007a]; HSEEP[2007b]).

Doyle *et al.*

Doyle *et al. Journal of Applied Volcanology* 2015 4:1, doi:10.1186/ s13617-014-0020-8

In NZ, MCDEM runs a voluntary participation National CDEM Exercise Programme which exercises at all levels of the CDEM structure listed above, where each tier exercise is informed by a "consistent regime of planning, observation, evaluation, feedback and continuous improvement" (MCDEM [2009a], p. 11). These are run within a 10 year schedule of exercise programming, with Tier 3 exercises every second year and Tier 4 exercises in the intervening years (MCDEM[2009b]). Individual organisations can participate in each of these exercises to the scale and scope they desire (e.g., ranging from a small scale in-house orientation exercise through to a multi-agency full scale response, Table 4).

Examples of recent Tier 4 exercises in NZ include the CDEM Exercise Tangaroa conducted in NZ to test the national response to a national tsunami warning in 2010 (MCDEM website [2014a]; Coetzee & Gale [2010]), and Exercise Ruaumoko which tested an all of nation response to a volcanic eruption in the Auckland Volcanic Field (discused earlier, MCDEM [2008]). An example of Tier 3 is the planned exercise Te Matau-a-Maui functional earthquake exercise to be run by the Hawke's Bay CDEM group to exercise the multi-organisational cross CDEM response to a MMVII scale earthquake in the region. An example of a Tier 2 is the (regional) Bay of Plenty CDEM group exercise of a severe weather event involving a storm surge leading to flooding and significant infrastructure damage in the region with a goal to exercise lifeline utility business continuity plans.

A full list of previous and planned exercises within the NZ CDEM sector can be found on the MCDEM website ([2014b]). These Tier 2, 3 and 4 exercises involve a wide range of co-ordination, collaboration, and considerable inter-organisational planning depending on the Tier level. In comparison, Tier 1 exercises involve just the individual organisation, such as a university exercising its response to a critical incident on campus (such as an earthquake or armed intruder). Similar tier structures are used by other organisations both nationally and internationally (e.g. FEMA's National Exercise Program, FEMA [2007]; FEMA website [2014]). For example, in NZ, Maritime New Zealand - Nō te rere moana Aotearoa (the Crown entity for Maritime safety, regulation, and emergency response) prepared contingency plans and runs exercises following three tiers depending on the level of responsibility: Tier 1 - industry (ships and onshore/offshore oil transfer sites), Tier 2 – regional councils, and Tier 3 – Maritime New Zealand.

Further to these national government led exercise schedules, there are a number of international collaboration programs to exercise response across nations. For example, Exercise Pacific Wave is run by the International Tsunami Information Center every two years to practice the sharing of information, warnings, advice, and resources while practicing government led decision making for a tsunami in the Pacific (UNESCO website [2014a]). Similar exercise schedules exist for the Indian Ocean and Caribbean Sea, as well as the Mediterranean (UNESCO website [2014b]). These provide useful examples of multi-agency international exercises, with a wide variety of organisations participating to various degrees (e.g. as watchers, partial participants, or full exercise participants). For example, 34 Pacific countries directly participated in Exercise Pacific Wave in 2008, and 10 additional countries 'watched' the exercise (i.e. received warning messages), which simulated the response to a tsunami induced by a M9.2 earthquake off northwest Japan (UNESCO website [2014c]). The purpose of this exercise was to evaluate and improve warning systems, and to improve the effectiveness of 'Member States' in responding to a destructive tsunami, including exercising of their operational lines of communications, warning systems, emergency response procedures and decision making, as well as to promote emergency preparedness. Exercises such as these could inform future multi-national global exercises in the volcanological community to enhance and inform best practice and protocols for international collaboration and response to an event that crosses country borders. This could be led by an international organisation like the International Association of Volcanology and Chemistry of the Earth's Interior (IAVCEI), to simulate and integrate the full scale global volcanological response to a *Black Swan*[b] volcanic event: such as a VEI 7 caldera eruption and the associated ash cloud, or managing the impacts of a fissure event like the 1783 Laki eruption.

[b]A *Black Swan* event (Taleb [2007]) describes those events that 1) exceed our expectations (an outlier), 2) have extreme impacts, and 3) are often 'rationalised by hindsight'. They are hard-to-predict, surprising rare events beyond "the realm of normal expectations", often characterised by high uncertainty. The probabilities of such events are often hard to compute due to the nature of the small probabilities involved. A recent natural hazard event which could be considered as a *Black Swan* event is the 2011 T hoku earthquake and tsunami and subsequent Fukushima nuclear power plant crisis; although some state this may not satisfy the first criteria (surprise) of a *Black Swan*(Lewis [2012]). The impact to aviation, travel, and trade during the 2010 Eyjafjallajökull eruption is another example of a *Black Swan* event.

The involvement of science advisors in these exercises is vital for the development of shared understanding and shared mental models between scientists and critical decision makers, as discussed above. For example, the Ruaumoko exercise in NZ (MCDEM [2008]; Smith [2009]; Lindsay et al. [2010]) involved science organisations (such as GNS Science) from exercise inception and scenario design, through to activation and response, and were used as opportunities to practice response, explore science advice, and try decision making tools relating to evacuation. However, it is vital that the scientists fulfilling the response role are rotated through the exercise, to represent realistic shift rotation in a prolonged event, and across exercises to ensure that as many potentially responding scientists as possible have the opportunity to practice these roles and build relationships with critical decision makers. Further, as discussed in Section 3, this rotation will enable performance management techniques to be practiced and refined, such as role swapping, shift rotation, rotation of the lead science agency or individual, and the balance of time and resources dedicated to the 'response role' and the 'day-to-day research role'. These become increasingly important during long duration responses providing resilience to the response (against untoward events, as well as contingent demands), while enabling a sharing of the load, management of competing or conflicting demands (both professional and personal) and maintenance of a fresh perspective via fatigue management.

Further, it is important that scientists, science agencies, and science advisory groups also lead and develop their own in-house and cross-science agency exercises to develop competencies around the science response itself. Thus we now review existing best practice guidelines for writing and evaluating such exercises.

Writing and Evaluating Exercises: Practice Guidelines

There are many guidelines existing in the emergency management and government literature that provide distinct steps to effectively write exercises (see Table 5). These all recommend the adoption of an exercise cycle, an example of which goes through the steps: 1) analysing the needs and outcomes desired for the exercise, which may be in line with process or performance improvement plans (this step may include a 'foundation' step assessing the status quo); 2) design and development of the exercise; 3) conducting the exercise; and 4) evaluating the exercise (see Figure 3).

Table 5: A sample of best practice guidelines available online for the design, writing and implementation of exercises (all websites last accessed on 23 April 2014)

Who	What & where
New Zealand Ministry of Civil Defence and Emergency Management	CDEM Exercises, Director's Guidelines for Civil Defence Emergency Management Groups [DGL 010/09]
	http://www.civildefence.govt.nz/assets/ Uploads/publications/dgl-10-09-cdem-exercises.pdf webcite
US Department of Homeland Security	Homeland Security Exercise and Evaluation Program has a range of documents available outlining effective exercise writing at: https://www.llis.dhs.gov/HSEEP webcite
US Department of Homeland Security, Federal Emergency Management Institute	FEMA's National preparedness directorate on National Training and Education runs a series of online courses across the range of Emergency Management at http://www.training.fema.gov/ webcite, which includes the unit IS-139 on 'Exercise Design' which contains a range of useful resources. http://training.fema.gov/emiweb/is/is139lst.aspwebcite
	See also the Ready.gov resources at: http://www.ready.gov/business/testing/exercises webcite
UK Cabinet Office & National Security and Intelligence	Emergency planning and Preparedness: exercises and training
	https://www.gov.uk/emergency-planning-and-preparedness-exercises-and-training webcite
	including 'The exercise planners guide': https://www.gov.uk/government/publications/the-exercise-planners-guidewebcite

National Directorate for Fire and Emergency Management, Ireland	A framework for major emergency management, Guidance Document 4: 'A guide to planning and staging exercises', May 2011.http://www.mem.ie/guidancedocuments/a%20guide%20to%20planning%20and%20staging%20exercises.pdfwebcite
Australian Emergency Management Institute	Managing Exercises, Handbook 3:http://www.em.gov.au/Publications/Australianemergencymanualseries/Documents/ManagingExercisesHandbook.PDFwebcite

Doyle *et al.*

Doyle *et al. Journal of Applied Volcanology* 2015 4:1, doi:10.1186/s13617-014-0020-8

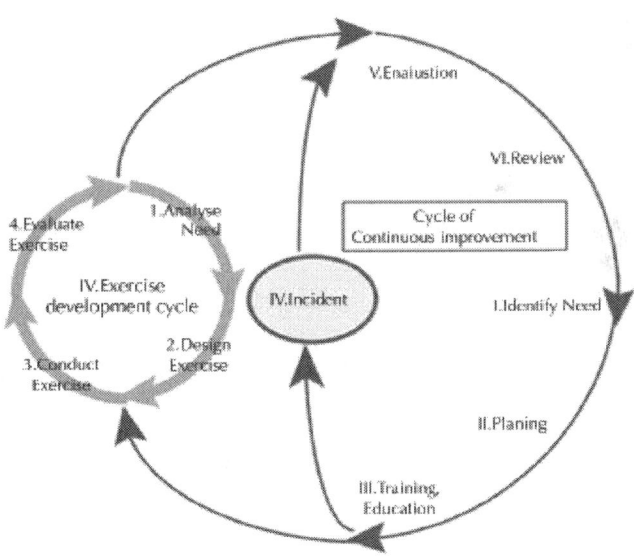

Figure 3: The cycle of continuous improvement and the exercise management model, after the Australian Emergency Management Institute (AEMI [2012]) and the Ministry of Civil Defence and Emergency Management (MCDEM [2009a]). Variants of this cycle are provided by each of the guidelines described in Table 5.

It is important that exercises do not just consider an event building up to a crescendo (e.g. the unrest period leading up to an eruption), but also consider non-events (the stalled eruption), response and recovery periods after an eruption (e.g. the period after a 'blue sky' eruption), or the issues of having a second volcanic crisis occurring alongside the focus volcano (e.g. in 2012 during the Te Maari unrest period in NZ, White Island also increased in activity requiring alert level changes). The volcanic events in NZ in 2012 provide a useful exercise scenario to examine the distribution of resources and personnel across two concurrent crises, and to identify more overt strategies for stress and fatigue management.

Exercises and simulations afford opportunities for agency representatives to develop, practice and review technical, management and team skills under realistic circumstances and to construct realistic performance expectations, and provide opportunities for developing creative problem solving competencies (within and between teams). They also provide opportunities to practice dealing with high pressure situations in a safe and supportive environment, increase awareness of stress reactions, and to rehearse strategies to minimize negative reactions (Flin [1996]). However, many exercises fail to mirror the complexity of disaster response environments (Paton and Auld [2006]).

There is a tendency sometimes to write these exercises like a 'play', providing too much structure and tailored or specific information. However, a truly challenging exercise, that provides the greatest opportunities for learning, should provide incomplete information, conflicting aspects, gaps in the situation description, ambiguous information, and unknowns; all creating an environment where participants are asking questions, not just finding solutions (Paton and Auld [2006]). For example, a monitoring team exercise could provide a situation in-exercise where one or many data feeds abruptly end. This provides participants a situation to address whether a machine or data relay has broken, or whether a significant event has occurred, allowing them to exercise the situation of uncertainty and potential decision and action strategies. Exercises should be designed in ways that allow individuals and teams to progressively push their boundaries and develop progressively more sophisticated competencies and relationships (Paton and Jackson [2002]). Only through exposure to exercises that challenge assumptions and allow personnel to confront novel events can preparation enhance response effectiveness for eruption events whose complexity will exceed even the most extreme simulation (Paton and Auld [2006]).

Thus these exercises should aim to reproduce reality as closely as possible, so decision makers and experts can experience the needs and realities of the advisory process in turbulent conditions (Borodzicz and van Haperen [2002]; Rosenthal and't Hart [1989]). However, this alleged realism can also be a danger as personnel may believe at the end of an exercise that they know what will happen in a real crisis (Borodzicz and van Haperen [2002]). Evaluation of exercises and events must be carefully conducted to minimise the risk of creating an optimistic bias that overestimates future response preparedness and capability, particularly if a real event has not constituted a major test of the response system (Paton et al. [1998a]). The development of an inter-agency exercise database would greatly assist this process.

Guidelines for evaluation and debriefings can be found in the emergency management literature (see Table 5 and Sinclair et al. [2012a]; Sinclair et al. [2012b]). However, these tend to focus on Key Performance Indicators (KPIs) and specific tasks, response plans, and communication process or outcomes and content evaluation. To ensure these exercise experiences truly enhance team response and development of a common understanding across the team, it is important to also consider team and group learning (Owen et al. [2013]), as well as the assessment of individual learning processes such as the progression from unconscious incompetence ('I don't know what I don't know') through to unconscious competence ('my skill has become second nature') (Conger and Mullen [1981]; Morell et al. [2002]; Thomson et al. [2006]). That is, an effective exercise is one that throws up several unanticipated issues (e.g., regarding information management, decision making, team functioning etc.) that can inform future training needs analyses and training and exercise design. Thus, exercising and scenario-based training should be regarded as an iterative process, where the debrief and evaluation is a vital part of that process to enhance the knowledge and skills of the participants, and to help them change their world view and individual paradigms, whilst also minimizing weaknesses caused by problems with exercise fidelity, such as a poor scenario or lack of reality (see review in Moats et al. [2008]).

Exercise Ruaumoko: the Importance of Exercises and Evaluation

Evidence for the benefits of exercises can be gleaned from Exercise Ruaumoko; benefits identified included opportunities to test the Auckland Volcanic Scientific Advisory Group (AVSAG, McDowell [2008]). It was a valuable learning experience for all concerned, illustrating the criticality of Auckland to the New Zealand economy and the vulnerability of key infrastructure elements (MCDEM [2008], p. 7). Further, the process of preparing for the Exercise motivated the formalisation of the AVSAG structure (McDowell [2008]) and helped to improve the understanding and performance of "individual elements" within a major CDEM operation (MCDEM [2008]). However, one of the greatest benefits of the exercise was that it highlighted the need for considerable work "to co-ordinate the various elements" of the inter-agency response into a "cohesive [future] response" (MCDEM[2008]). This exercise thus allowed responding agencies to identify response, network and communication issues that need to be modified for effective response and the importance of testing assumptions under realistic crisis conditions, such as high stress, short time, and low resource conditions with competing demands, to demonstrate faults or potential faults in the system. We thus now further discuss Exercise Ruaumoko in detail to provide an example of an exercise structure and to illustrate the benefits of effective evaluation.

Exercise Ruaumoko was a MCDEM led exercise, with over 1,500 participants from approximately 125 participating organisations (MCDEM [2008], p. 8). It was the second in a series of ongoing "All of Nation" exercises led by MCDEM (MCDEM website [2014c]). Exercise Ruaumoko also provided a focus to develop and formalise the AVSAG structure (which was co-ordinated and arranged by Auckland CDEM), while also implementing Terms of Reference, and arranging formal contract agreements for the participating scientists (Cronin [2008]); and to test such processes within a scenario. The three core objectives of the exercise were to: 1) understand, develop, and practice the respective roles and responsibilities of local, regional, and national agencies in response to the exercise scenario (McDowell [2008], p. 7; MCDEM [2008], p. 5); 2) embed the planning arrangements in standard processes for all participating agencies; and 3) confirm the connections between local, regional, national and international agencies. In addition, it enabled staff at the key levels to "practice and develop planning and the management of response activities using the connections and procedures that are in place", and also practice the roles, responsibilities and procedures that are included in the National CDEM plan (MCDEM [2008], p. 55). The key sub-objectives of the exercise most relevant to science advice included the examination and testing of the coordination of science aspects and the management of public information and education. While Exercise Ruaumoko was primarily written and co-ordinated by MCDEM with Civil Defence objectives, it still provides a useful example for volcanologists of a large scale volcanology exercise, and the monitoring information and science response parts could be "re-exercised" or used as inspiration for a smaller scale science led exercise.

One of the unique aspects of Exercise Ruaumoko was its timeline, which involved periods of "unfelt earthquakes" in November 2007 and then again between 3rd to 7th March 2008. This was followed by a week of increased activity and full exercise play from Monday 10th to Friday 14th March, 2008 (MCDEM [2008], p. 11; see Table 6). Throughout this scenario, GNS Science and its duty volcanology team received raw geophysical data such that they could directly exercise their analysis and interpretation processes, participate within the AVSAG response, practice their own decision making, and produce Science Alert Bulletins, which in turn formed injects for the rest of the exercise participants (CDEM and other response and welfare agencies). The scenario focused on the lead-in to a volcanic eruption, stopping shortly after the eruption started (MCDEM [2008], p. 8), thus the concept of the exercise was to provide for pre-emergency planning rather than operating in the traditional post-event response only mode often typified of exercises (MCDEM [2008], p. 9). Detailed information regarding the exercise management and planning process, which started in 2006, can be found in MCDEM ([2008]). Great lengths were undertaken to ensure this exercise was realistic, with a scientist "acting the volcano" and providing monitoring data injects, and the eventual outcome (including vent location in this monogenetic field) being kept secret from all participants. The wide number of participating organisations added realism in terms of inter-organisational communication and networks (Section 3). However, realism in this exercise was reduced by AVSAG and the GeoNet "actors" not being expected to participate in the exercise outside normal business hours. On the night of Thursday 13th, this became a particular issue as the emergency managers and decision makers continued to conduct exercise "play" throughout the night in the absence of the science advisors.

Table 6: The scenario timeline of Exercise Ruaumoko (MCDEM [2008])

November 2007	• Initial phase of the exercise: initial earthquakes characterised as 'deep long period earthquakes' and suspected by seismologists participating in the exercise to be of volcanic origin. Due to the rarity of such earthquakes in the Auckland region, the GNS Science/GeoNet exercise participants raised the Scientific Alert Level for the Auckland Volcanic Field to 1 ("Initial signs of volcanic unrest")*.
	• Fourty eight hours later this seismic activity appeared to stop, with no observable shallowing of earthquake depths. The exercise scientists determined after a week of quiescence that "if there had been an intrusion of magma, it had likely stopped or 'failed' at a depth of 40–50 km" (MCDEM, p. 11).
	• Scientific Alert level reduced to 0 ("Usual dormant or quiescent state") on 16th November
March 2008	• Second phase of the exercise: seismicity resumed and became sustained causing concern for the 'exercising' authorities and Auckland communities, as the seismic source progressively shallowed, indicating that a volcanic eruption may occur. However, the geographic distribution of earthquake epicentres indicated a few possible locations for such an eruption in the monogenetic field.
3rdMarch	• Scientific Alert Level raised to Level 1 ("Initial signs of volcanic unrest")
8thMarch	• Scientific Alert Level raised to Level 2 ("Confirmation of volcano unrest")

9th-12thMarch	• Earthquakes started to be 'felt' across the Auckland region as the seismic source reached 20–25 km depth, and by 20 km more than 50 were being recorded a day.
12thMarch	• At 10–15 km depth the events exceeded 100 per day, and by 12th March the earthquakes started to cluster in one region (Mt Roskill-Hillsborough to Mangere), such that ACDEM started to draw up evacuation maps.
12thMarch	• Scientific Alert Level raised to Level 3 ("Real possibility of Hazardous eruptions")
13thMarch	• Clustering of earthquakes refined further (Mangere Inlet) with 300 events 'recorded' in a 24 hour period with depths of 5 km and many being of Mercalli 4.5 and 'felt' by residents.
14thMarch	• By the morning of Friday 14th March, earthquake merged into tremor, increasing in strength throughout the morning with ground cracking and slumping observed around the Kiwi Esplanade and Mangere Bridge, and by 1.50 pm a small phreatomagmatic eruption had begun.
14thMarch	• Scientific Alert Level raised to Level 4 ("Hazardous local eruption in progress")
	• End Of Exercise

Doyle *et al.*

Doyle *et al.* *Journal of Applied Volcanology* 2015 4:1, doi:10.1186/s13617-014-0020-8

Cognisant of this issue, there are still many lessons that were identified from exercise reviews that have direct implications for the improvement of future response. Exercise evaluations we consider here include: the in-house GNS Science review of the AVSAG and GeoNet process, a NZ volcanologist who acted as an observer to the exercise including all AVSAG meetings and conference calls (Cronin [2008]), MCDEM's all of exercise evaluation (MCDEM [2008], and Table 7), and Auckland's' Civil Defence Emergency Management Group evaluation report (McDowell [2008]). Several key issues were observed and lessons identified which we now discuss in reference to our understanding from Section 3 of situational awareness, mental models, trust and decision making for effective response, as follows:

The AVSAG was constructed of a tri-partite sub-group system (Volcano Monitoring, Volcanology & Social Consequence, see Section 2.1), which by its nature had GeoNet (the monitoring arm of GNS Science, legislated to provide official advice) sitting as a separate sub-group to the volcanology sub-group which encompassed university and other Crown Research Institutes, Science agencies (e.g. MetService) and GNS Science; with meetings between these groups conducted via regular teleconferences. An extremely positive aspect of the AVSAG structure was its inclusiveness (McDowell [2008]), however it was seen as being too "cumbersome during the escalation of events in the scenario" (Cronin [2008], p. 6). In the early stages of the exercises, information flow between these two sub-groups was adequate. However, as the crisis progressed, information transfer reduced considerably and there was poor communication (Cronin [2008]; McDowell [2008]). This represents a significant issue for effective individual and group situational awareness and would have impacted upon shared mental models of the situation, resources required, and decisions to be made (Section 3).

AVSAG tasks were driven by questions provided by the Auckland EOC and the National Crisis Management Centre. As stated by Cronin ([2008]), very little additional advice was provided, even though some questions were not relevant. At times the questions were superseded by events, and by concentrating on these questions participants felt they "had excluded additional important comments" (Cronin [2008], p. 5). The exception to this was the social sub-group, who did pre-empt questions. When we consider this in light of our discussions in Section 3, we can see that this information sharing is typical of *explicit* information requests and represents a weak shared mental model of the issues and information needs, much like the *ad-hoc* communication characterised in the response to Ruapehu 1995–1996 (Section 3.3, & Paton et al. [1998a]). This demonstrates a need to conduct further activities to develop shared mental models further, and strengthen inter-organisation networks and communication.

Due process outlined in AVSAG was followed in situations when *novel* approaches would have been more appropriate (Sections 3.1

and 3.3). For example, as the exercise reached crisis, AVSAG followed process by only responding to explicit requests from the decision makers (Cronin [2008]), to the point that events actually superseded their responses. Failure to recognise the implications of future status on the changing need for information represents a failure of situational awareness (in terms of future status), and weak shared mental models (understanding other demands on the recipient; Section 3). The in-house GNS Science Review also cautioned that very structured meetings could constrain science discussion, which would thereby limit full debate and exploration of novel considerations and more creative decision making processes. This highlights how decision making undertaken within normal scientific research (long time for decisions, considerable debate, more *analytical* decision processes) differs from that in emergency management environments (rapid decisions for public safety in high risk/low time contexts which call for more *recognition primed and intuitive* decision processes). It was thus felt that the AVSAG process was good for 'peace-time' but unwieldy "when rapid response is required".

At the crisis point of the exercise, as evacuation planning occurred, AVSAG was "side-stepped" despite there "being new scientific information to hand about the eruption location and possible eruption styles" (Cronin [2008], p. 4). This was a direct reflection of the tri-partite structure of AVSAG failing at the heat of the crisis, and GeoNet carrying out two roles: 1) as the monitoring arm within AVSAG, and 2) being mandated to be the main information provider to government, the latter of which took priority. The escalation of the crisis resulted in a perception that time was too short for wider deliberation (Cronin [2008]; p. 5), and resulted in a focus on the rapid assessment and decision making in relation to technical data alone (McDowell [2008]). This however resulted in a lack of integration across the wider volcanological community represented in the volcanology sub-group of AVSAG, and thereby a breakdown of shared situational awareness (Section 3.2) and an artificial division between these scientists (McDowell [2008]).

The role of liaison officers (GeoNet duty officers) at the Auckland Group EOC and NCMC was vital for communication, and information sharing, (see Sections 3 and 4). Auckland Civil Defence and Emergency Management Group (ACDEM) (McDowell [2008]) identified that it was highly beneficial to have a scientist in the GEOC as they provided instant integration of science advice into their decision making. The presence of the liaison officers would have enhanced the situational awareness of the emergency managers, and communication within the GEOC most likely moved towards an implicit supply of information by the liaison officer as they recognised a need for their advice. However, as discussed in Section 2.1, as the crisis progressed the information delivered by the two liaison officers diverged, an artefact of the breakdown of shared situational awareness between their two dispersed geographical locations as the time and risk pressures increased. As stated by Cronin ([2008]), p. 7, with this approach "there is no guarantee that the two EOC's will receive the same advice", and in the exercise this resulted in actions by the NCMC that were a surprise to the Group EOC. This demonstrates the importance of consistent science advice due to the impact any divergence may have upon cohesive integrated emergency management decision making and response. From this, the reviews identified a need for policy to be developed for periods of extreme urgency to ensure "only one representative communicates the same message to National and Group controllers".

Reviews also identified a need for clarity around probabilities and uncertainty, and that there had been a use of geological terminology and phraseology that was not understood by emergency management, resulting in the recommendation to identify what the emergency managers need and how they use this information (see also Doyle et al. [2014a], [b]). As stated by MCDEM ([2008]), p. 50, "there is a need for further analysis and/or translation of primary science information to limit misinterpretation and to ensure it meets the needs of all organisations".

Communication infrastructure issues were also noticed, and a need for clarity around communication and reporting channels, as well as ensuring documentation was readily available for participants not able to attend meetings. This is vital for maintaining good shared situational awareness of the evolving crisis.

Work load was identified as an issue for responding scientists due to their dual roles (both response and day-to-day roles) as well as the multitasking demands placed upon them within the EOC. This indicates a need to consider the use of multiple liaison officers within an EOC, to share these demands, while both assisting with the maintenance of situational awareness within the EOC regarding developments of the science and data inputs. However, messages must be consistent when there are multiple liaison officers within an EOC, to prevent the potential divergence of advice discussed above.

Table7: Recommendation 14 "Understanding Volcanic Hazard and Communicating Science Advice", from the 17 recommendations suggested by MCDEM after Exercise Ruaumoko (MCDEM [2008], p. 50–51)

... there is a need for further analysis and/or translation of primary science information to limit misinterpretation and to ensure it meets the needs of all organisations.	
... there is a need to review local and national requirements for science advice (in planning and response periods), and look at processes for achieving better co-ordination and synchronisation, while also meeting individual agency requirements.	
Recommendation 14:	**MCDEM** should lead work with **CDEM Groups** and **science agencies** to:
	14.1 Consider options for integrating local and national science capabilities and processes
	14.2 Facilitate collaborative planning by science agencies, including universities, for post-event science investigations
	CDEM and science agencies should:
	14.3 Champion collaborative public-good research to enhance the scientific understanding of the Auckland volcanic system, in particular its precursor and eruptive behaviours
	14.4 Support on-going volcanic hazard education about the extent, size and nature of hazards and impacts.

Doyle *et al.*

Doyle *et al. Journal of Applied Volcanology* 2015 4:1, doi: 10.1186/s13617-014-0020-8

From these observations, and this comprehensive testing of AVSAG, specific recommendations for improving the science advice provision and response included (MCDEM [2008]; McDowell [2008]; Cronin [2008]; Table 7):

The need to investigate the options for a merged volcanology-subgroup encompassing the GNS Science/GeoNet monitoring arm in such a way that the overall group is not too cumbersome, but such that all expertise can be involved in the response.

Recognising and ensuring that both the volcanology and monitoring sub-groups of AVSAG have access to the same data resources *(which we also note would help improve shared situational awareness)*.

That MCDEM must work with CDEM and Science agencies to 1) consider options for integrating local and national science capabilities and processes, 2) facilitate collaborative planning by science agencies, including universities, for post-event science investigations (Smith [2009], p. 76) while also meeting individual agency requirements (MCDEM [2008], p. 50).

Adopting a structure where GeoNet (which has an always active on-call duty team) has the main advisory role, whilst "co-opting external AVSAG scientists and the GeoNet team into a volcanology/monitoring SAG" as needed (Cronin [2008], p. 9). This aims to ensure the wider NZ expertise can contribute to the understanding of the situation and incorporates locally based scientists into the response process.

These observations clearly illustrate the benefits of exercises that truly test scientific response, VSAG structures, and inter-organisation communication and decision-making responses. The 'split' observed between the monitoring and wider volcanology subgroup of AVSAG is interesting to consider in the context of the influence of group identity (Hogg [1992]) and the 'stereotyping of in- and out- groups' that can affect collaboration in a response (Paton et al. [1998a]) and reduce the capacity to develop swift trust.

As briefly introduced in Section 2, in response to these exercise evaluations, Smith ([2009]) (on behalf of MCDEM), outlined a potential alternative model provided by a NZ Volcanic Advisory Panel (NZVAP), to avoid such a breakdown in communication, as well as to integrate local and national science research response (Figure 2b). This model was developed through a MCDEM facilitated dialogue between volcanologists to identify how the skills of scientists in the range of different organisations (including universities, Crown Research Institutes, consultancies and councils) can be integrated "in support of a national science agency such as GNS Science that has responsibilities under the National CDEM Plan for providing warnings and advice" (Smith [2009], p. 76). This national advisory model aims to mobilise science capability while *remaining responsive to local CDEM needs* (ibid, p. 77), and the intention of such a national mechanism is not to override any existing scientific or planning advisory groups at the local level, but rather that it would be complementary and "provide for a level of consistency for how New Zealand-wide science capability is mobilised when a large-scale science response is needed". Other benefits of such a group include: a) facilitating coordinated post-event investigations and data sharing arrangements, b) providing strategic advice on research direction and priorities, c) fostering connectivity across the physical and social sciences, and d) supporting alignment between researchers and research users (see also Jolly and Smith [2012]).

This proposed model was still under consideration and discussion when the Canterbury earthquake sequence started in September 2010. This sequence required an extensive response of many of the same scientists and agencies, and thus the development and implementation of such a NZVAP model has been put on hold until extensive reviews of the inter-agency response to those earthquakes were conducted (including both a Royal Commission of Inquiry and a Coronial Inquiry) and until the science advisory process during those earthquakes, and the more recent volcanic eruptions of Te Maari in 2012, have been fully evaluated by agencies and through various research projects in process (see also Section 2). However, until those real event reviews are available, Exercise Ruaumoko represents a clear case study of how exercises can be used to test a new VSAG structure, identify issues in that structure and pave the way for the development of a new structure. As stated by MCDEM ([2008]), p. 55 "there were major achievements in the preparatory phase [of Exercise Ruaumoko] that simply wouldn't have occurred without the context of the exercise". After formalisation of any new NZVAP structure, the next step should be test this through another exercise (of a different scenario) to complete the exercise cycle (Figure 3) and implement any lessons from that as part of a regular evaluation of science advisory mechanisms.

DISCUSSION: A NEW EXERCISE STRUCTURE FOR VOLCANOLOGY

It is important that science agencies both participate in and lead exercises of all scales and types (Table 4) due to the many benefits discussed above. In particular, benefits can arise from smaller exercises that may be more suitable for restricted budgets. For example a full-scale exercise the scale of Ruaumoko would require significant exercise resources in terms of both personnel and costs. For that exercise, GNS Science contributed both exercise writing and preparation time of approximately 2 weeks for 2 people, and in the exercise itself approximately 18 hours of meeting time (plus associated data analysis time) for 10 people. Thus, in total, approximately 300 hours of professional time were contributed towards exercise development, exercise play, and evaluations and debriefs (detailed costs were not available, thus these are estimates only). This equates to approximately 70,000-100,000 NZD for a science agency to participate in a full scale CDEM led national exercise like Ruaumoko with realistic timelines (including a full week of exercise play). Equivalent costs for MCDEM (as exercise writing management team, as well as in response) would be in the range of 1–1.75 full time equivalent (FTE) staff members for 12 months of exercise preparation, escalating to 3FTE for the last 3 to 4 months and then full team engagement in the exercise play itself. These FTE staffing costs would be in addition to operational expenses (in excess of approximately 200,000 NZD) for workshops, briefings, travel, report writing, and public education material. Meanwhile for regional/local CDEM, typical costs for participating in such an exercise would depend on their exercise involvement, ranging from 500,000 NZD for full scale participation in a Tier 4 exercise (Auckland CDEM during Exercise Ruaumoko) through to 5000–10,000 NZD for smaller Tier 1 and Tier 2 exercises (e.g. Wellington Regional Emergency Management Office, based on regular Exercise Phoenix series). Thus it is important that agencies consider conducting a cost-benefit analysis when deciding the scale of an exercise, being aware that benefits from conducting exercises include the considerable reduction of personnel and financial costs during a real response due to the increased capability developed in the exercise.

It is clear from the above that full scale participation in CDEM led national exercises, while extremely beneficial for the future response capability of scientists and volcanologists, can be very expensive in terms of resources and professional time. It is thus important that scientists do not just participate in CDEM led exercises, practicing in liaison roles and on expert panels, but also exercise their own agency and inter-agency response as well. By exercising within their own agencies, scientists can enhance intra- and inter-agency collaborations. By developing, writing and leading an exercise, scientific agencies can design exercises to meet their core goals in terms of capability building and competency testing; to tailor them to meet their needs for skills and process development; to meet their available budget by providing opportunities to develop smaller, less costly exercises to test processes within just a small team (e.g. a monitoring team could practice their response, analysis and decision making responses). These smaller exercises are still greatly beneficial to response capability and to improved shared mental models of the response environment within that team. However they currently occur in an ad hoc manner, with no co-ordinated Tier structure like that of the CDEM sector (Section 5). We thus suggest there is a need to formalise such a structure for volcanology to help promote a regular exercise cycle, and to help integrate such processes into CDEM frameworks, as well as to encourage documentation of such exercises for future shared use, and propose an equivalent scale as follows:

Level 1 – in-House: within-Team or within-Group Exercise

E.g. exercise the response of a specific team, such as an agency monitoring team or a university research group, to an event such as a lahar breakout and the associated opportunity for the rapid collection and dissemination of data. This could involve story boarding the response and requirements through a facilitated table top exercise, through to a functional exercise or practice run involving mobilisation, data collection procedures and radio communication tests.

Level 2 – in-House: within-Agency Exercise

E.g. exercise the agency response to an eruptive episode, including the various responding monitoring teams within this organisation, such as: geochemistry, geophysics, geodesy, and social sciences, as well as those responsible for issuing monitoring statements, developing forecasts, and providing impact and warning advice to the public and key decision makers. Through this, an organisation can develop an understanding of the dependencies that each of these teams or groups have upon each other and their shared requirements to perform their role, as well as developing relationships across these roles. In addition, abilities to continue the agency's day-to-day duties alongside this response can be explored and issues identified. Examples of such an exercise range from a small facilitated table top exercise to practice a monitoring meeting and the resultant change of an alert level, through to an inject led functional exercise utilising monitoring injects throughout a prolonged period of time.

Level 3 – Regional or National: Across-agency Exercise

E.g. exercise members of a national volcanic science advisory group, for a long duration volcanic unrest period. This advisory group would bring together individuals from across universities, agencies and technical organisations, as well as any nationwide monitoring agency, and thus it provides an opportunity to exercise both their roles and others within their agencies and organisations who may support them in their role. For example, exercising a volcanic ash fall event provides an opportunity to practice interaction across agencies from the volcanological, meteorological, public health, veterinary and farming sectors, and beyond.

Level 4 – International: Across-agency Exercise

E.g. exercise the response to large scale ash cloud hazard event. In a manner similar to Pacific Wave, this could involve members ranging from the local weather monitoring and forecast organisation, through to local and international volcanologists, meteorologists, and aviation authorities. Through this the international sharing of information necessary for local, regional and national ash cloud monitoring and forecast can be exercised, as well as collaboration in the preparation and delivery of any aviation warnings.

Each of these exercise levels could involve only the scientific agency, or other civil defence and responding organisations, and may range from orientation to full scale exercises (see Table 4), as appropriate. We use the term 'Level' to avoid confusion with the term 'Tier' used in the emergency management led exercises. These exercise levels may feed in to any of the CDEM Tier 1 to 4 scales, as desired by the co-ordinating scientific agency. For example, a scientific agency may choose to only exercise their landslide assessment team (Level 1) as part of a Tier 3 CDEM earthquake exercise, or they could choose to exercise all their responding teams to this scenario, including, for example, a full inter-agency tsunami expert panel (Level 3). The key difference for our proposed exercise structure is that the exercise planning and writing is led by the scientific agencies, and not a local, regional or national civil defence agency, and thus is directed by the core goals and competencies that the scientific agency wishes to develop.

Further to these four levels, we also suggest that volcanologists exercise with a different hazard. For example, a team of volcanologists could run 1) a table top flood exercise themselves (with the scenario provided by the appropriate agency) 'acting' as flood scientists; or 2) watch or participate in a full scale flood exercise, again playing the role of flood scientists. By exercising an 'unfamiliar hazard' (e.g. the rapid, high frequency, sudden onset flood), volcanologists cannot make their usual assumptions, and so the process can help to identify issues and unrealistic expectations of response capability. It also gives these volcanologists an opportunity to learn techniques, procedures and 'tips' from these other scientific communities that may be more experienced in emergency response roles. For example, some flood and weather agencies contain scientists who are very practiced at regularly switching from their 'research' role to a 'response' role, forming rapid response teams, and working alongside CDEM and local governments on a regular basis.

CONCLUDING REMARKS

Through a review of the literature on emergency management team response, decision making, mental models, situational awareness and exercising, particularly in the NZ context, we argue that to truly enhance the science response during a disaster, science agencies, and science advisory groups, must learn from emergency management agencies and embark on a suite of training activities including exercise and simulation programs within their own organisations, rather than solely participating as external players in emergency management activities. Through this in-house training, the future response capability of science advisory groups and agencies can be enhanced, while also developing their true understanding of the needs of the emergency managers and other key decision makers. This enhanced understanding may lead to a modification of existing scientific programmes and projects, with a view to enhance outputs to meet decision makers' needs in advance of a crisis rather than during the crisis itself. Further, by participating in these activities we can enhance the response capability by developing shared mental models across the team about the response issues, demands, tasks and interpositional knowledge for members within and across teams.

These exercises not only provide opportunities to practice communications and plans and enhance these team mental models, but also for scientific agencies to rehearse strategies to convey uncertainty and how it could be included (Doyle and Johnston [2011]; Doyle et al. [2014a] & Doyle et al. [2014b]), and for other agencies to be able to develop associated contingent planning. In addition, exercise and scenario planning can also help to answer if we are doing the right science for the response as well as for the advancement of science. There is thus a necessity for organisations to provide structure, resources and time to help facilitate and promote such training activities, as well as to encourage performance management techniques within a response, such as role swapping, while also acknowledging our responsibility as scientists to fulfil this response role effectively.

Beyond exercises for response it is also important to acknowledge that such exercises can advance our scientific knowledge and strategies for response performance, learning, and decision competencies. Volcanic scenarios provide an important case study for these research exercises into human behaviour, learning and communication, due to their highly uncertain nature in terms of expected scale, scope, timing and impacts (e.g. Doyle et al. [2011]; Dohaney [2013]).

There is also a need, through such research exercises, to identify ways to develop 'swift trust' in an event. Trust is intrinsic to shared mental models, distributed decision making, and information sharing in a response, particularly during situations of high uncertainty. Identifying strategies to create swift trust goes beyond volcanic response management, and has implications for developing effective multi-agency collaboration when agency representatives find themselves managing any hazard crises without prior contact. For information sharing it also has implications as it moves people from an individualistic perspective ('my information is best') to an acknowledgement of the collective perspective and role ('my information is one piece in the big picture of response').

AUTHORS' CONTRIBUTIONS

EEHD conceived the design of the manuscript and the literature review, and lead the draft of the manuscript. DP contributed to the design and helped draft the manuscript. DMJ helped draft the manuscript. All authors read and approved the final manuscript.

ACKNOWLEDGMENTS

EEHD was supported by a Foundation for Research Science & Technology NZ S&T Postdoctoral Fellowship MAUX0910; and thanks Sally Potter and Sarah Beaven for many discussions that have helped form thinking around exercising in volcanology; Vince Neall and Bob Stewart for sharing information about the early days of the Egmont Volcanic Advisory Group (now Taranaki Seismic and Volcanic Science Advisory Group); Brad Scott, Harry Keys and Gill Jolly for sharing information about the other Volcanic Advisory Groups in NZ; Richard Smith and MCDEM for the use of Figure 2a and b; and also various agencies for providing approximate costs of exercise participation. We also thank three anonymous reviewers and the Guest Editor, Jan Lindsay, for comments and suggestions that have greatly enhanced this manuscript.

REFERENCES

1.   (2012) Managing Exercises, Australian Emergency Management Handbook Series, Handbook 3, Second. 128.

2.   Barclay J, Haynes K, Mitchell T, Solana C, Teeuw R, Darnell A, Crosweller HS, Cole P, Pyle DM, Lowe C, Fearnley C, Kelman I (2008) Framing volcanic risk communication within disaster risk reduction: finding ways for the social and physical sciences to work together. Geol Soc London, Spec Publ 305:163-177 doi:10.1144/SP305.14

3.   Bayley S (2004) Living with Volcanoes: The Taranaki Story. Tephra. Ministry of Civil Defence & Emergency Management, Wellington, NZ.

4.   (2002) Contingency Plan for the Auckland Volcanic Field.

5.   Ben-Shalom U, Lehrer Z, Ben-Ari E (2005) Cohesion during military operations: a field study on combat units in the al-Aqsa intifada. Armed Forces Soc 32(1):63-79

6.   Blickensderfer E, Cannon-Bowers JA, Salas E (1998) Cross-training and team performance. In: Cannon-Bowers JA, Salas E (eds) Making Decisions Under Stress: Implications for Individual and Tream Training, American Psychological Association, Washington, D.C., USA. pp 299-311

7.   Bloom M, Menefee MK (2014) Scenario planning contingency planning. Public Product Manag Rev 17:223-230

8.   Boin A, 't Hart P (2001) Public leadership in times of crisis: mission impossible? Public Adm Rev 63:544-553

9.   Borodzicz E, van Haperen K (2002) Individual and group learning in crisis simulations. J Contingencies Cris Manag 10:139-147 doi:10.1111/1468-5973.00190

10.   Cannon-Bowers JA, Bell HE (1997) Training decision makers for complex environments: implications of the naturalistic decision making perspective. In: Zsambok CE, Klein G (eds) Naturalistic Decision Making, Lawrence Erlbaum Associates, Mahwah, NJ. pp 99-110

11.   Cannon-Bowers J, Salas E, Blickensderfer E, Bowers C (1998) The impact of cross-training and workload on team functioning: a

replication and extension of initial findings. Hum Factors J Hum Factors Ergon Soc 40:92-101 doi:10.1518/001872098779480550

12. Canterbury Earthquakes Royal Commission – Te Komihana Rūwhenua o Waitaha (2012) [http://canterbury.royalcommission. govt.nz/], accessed 23 April 2014

13. Coetzee D, Gale N (2010) New Zealand Exercise "Exercise Tangaroa" [http://www.ioc-tsunami.org/index.php?option=com_ content&view=article&id=59:new-zealand-tsunami-exercise-exercise-tangaroa], accessed 9th December 2014

14. Conger DS, Mullen D (1981) Life skills. Int Jounral Adv Couns 319:305-319

15. Coppola NW, Hiltz SR, Rotter NG (2004) Building trust in virtual teams. IEEE Trans Prof Comm 47(2):95-104

16. CPVAG (2009) Central Plateau Volcanic Advisory Group Strategy, October 2009. Report No 2010/EXT/1117. Morris B (eds) Horizons Regional Council, Palmerston North, NZ. https://www.horizons. govt.nz/assets/publications/keeping-people-safe-publications/ Central-Plateau-Volcanic-Advisory-Group-Volcanic-Strategy. pdf. Accessed 9 December 2014

17. Crego J, Spinks T (1997) Critical Incident management simulation. In: Flin R, Salas E, Strub M, Martin L (eds) Decision Making Under Stress Emerging Themes and Applications, Ashgate Publishing Limited, Aldershot, England. pp 85-94

18. Crichton M, Flin R (2002) Command decision making. In: Flin R, Arbuthnot K (eds) Incident Command: Tales from the Hot Seat, Ashgate Publishing Limited, Aldershot, England. pp 201-238

19. Crisp CB, Jarvenpaa SL (2013) Swift trust in global virtual teams. J Personnel Psychol 12(1):45-56

20. Cronin SJ (2008) The Auckland Volcano Scientific Advisory Group during Exercise Ruaumoko: observations and recommendations Civ. Def. Emerg. Manag. Exerc: Ruaumoko. Auckland Regional Council, Auckland.

21. Curnin S, Owen C, Brooks B, Paton D (2015) A theoretical framework for negotiating the path of emergency management multi-agency coordination. Appl Ergon 47:300-307

22. Dirks K, Ferrin D (2001) The role of trust in organizational settings. Organ Sci 12:450-467

23. Dohaney JAM (2013) Educational Theory & Practice For Skill Development In The Geosciences. PhD Dissertation, University of Canterbury, Geological Sciences, NZ.

24. Doyle EE, Johnston DM (2011) Science advice for critical decision-making. In: Paton D, Violanti J (eds) Working in High Risk Environments: Developing Sustained Resilience, Charles C Thomas, Springfield, Illinois, USA. pp 69-92

25. Doyle EEH, Johnston D, Paton D (2011) Investigating science advice, emergency management and decision making in the laboratory. XXV IUGG Gen. Assem. International Union of Geodesy and Geophysics, Melbourne, VIC, Aust.

26. Doyle EEH, McClure J, Johnston DM, Paton D (2014) Communicating likelihoods and probabilities in forecasts of volcanic eruptions. J Volcanol Geotherm Res 272:1-15, doi:10.1016/j.jvolgeores.2013.12.006

27. Doyle EEH, McClure J, Paton D, Johnston DM (2014) Uncertainty and decision making: volcanic crisis scenarios. Int J Disaster Risk Reduct 10:75-101

28. Endsley MR (1994) Situation awareness in dynamic human decision making: theory. In: Gilson RD, Garland DJ, Koonce JM (eds) Situational Awreness in Complex Systems: Proceedings of a Cahfa Conference, Embry-Riddle Aeronautical University Press, Daytona Beach, FL. pp 27-58

29. Endsley MR (1997) The role of situation awareness in naturalistic decision making. In: Zsambok CE, Klein G (eds) Naturalistic Decision Making, Lawrence Erlbaum Associates, Mahwah, NJ. pp 269-284

30. EQC Website (2014) The Earthquake Commission [http://www.eqc.govt.nz/about-eqc], accessed 17th September 2014

31. FEMA (2007) National Exercise Program (NEP) US Department of Homeland Security, March 8 2007, WJTSC 07–1.

32. FEMA website (2014) National Exercise Program (NEP), FEMA, US Department of Homeland Security [http://www.fema.gov/national-exercise-program], accessed 23 April 2014

33. Flin R (1996) Sitting in the hot seat: Leaders and Teams for Critical Incident Management. John Wiley & Sons, Ltd, Chichester, England.

34. Ford JK, Schmidt AM (2000) Emergency response training: strategies for enhancing real-world performance. J Hazard Mater 75:195-215

35. Funtowicz SO, Ravetz JR (1991) A new scientific methodology for global environmental issues. In: Costanza R (ed) Ecological Economics: The Science and Management of Sustainability, Columbia University Press, New York. pp 137-152

36. Funtowicz SO, Ravetz JR (2003) Post-normal Science. In: International Society for Ecological Economics (ed) Online Encyclopedia of Ecological Economics. Last accessed 8th Dec 2014 from http://isecoeco.org/pdf/pstnormsc.pdf

37. GeoNet website (2014) GeoNet: About us. [http://info.geonet.org.nz/display/geonet/About+GeoNet], accessed 17th September 2014

38. GNS Science website (2014) GNS Science: About us. [http://www.gns.cri.nz/Home/About-Us], accessed 17th September 2014

39. Goodman R, Goodman L (1976) Some management issues in temporary systems: a study of professional development and manpower-the theater case. Administrative Sci Q 21(3):494-501

40. Greenbaum CW (1979) The small group under the gun: Uses of small groups in battle conditions. J Appl Behav Sci 15:392-405

41. Bammer G, Smithson M (eds) (2008) Emergency management thrives on uncertainty. In: Uncertainty and Risk Multidisciplinary Perspectives Earthscan, London, UK. pp 231-243

42. Hogg MA (1992) The Social Psychology of Group Cohesiveness: From Attraction to Social Identity. New York University Press, New York, NY.

43. (2007) Homeland Security Exercise and Evaluation Program (HSEEP). In: HSEEP Overview and Exercise Program Management. US Department of Homeland Security, Washington, DC. p 82

44. (2007) Homeland Security Exercise and Evaluation Program (HSEEP). In: Volume II: Exercise Planning and Conduct. US Department of Homeland Security, Washington, DC. p 57

45. Hyllengren P, Larsson G, Fors M, Sjöberg M, Eid J, Olsen OK (2011) Swift trust in leaders in temporary military groups. Team Perform Manage 17(7/8):354-368

46. (1999) IAVCEI subcomittee for crisis protocols. Professional conduct of scientists during volcanic crises. Bull Volcanol 60:323-334

47. Janssen M, Lee J, Bharosa N, Cresswell A (2010) Advances in multi-agency disaster management: Key elements in disaster research. Inf Syst Front 12:1-7 doi:10.1007/s10796-009-9176-x

48. Johnston DM, Houghton BF, Neall VE, Ronan KR, Paton D (2000) Impacts of the 1945 and 1995–1996 Ruaupehu eruptions, New Zealand: an example of increasing societal vulnerability. Geol Soc Am Bull 112:720-726 doi:10.1130/0016-7606(2000)112<720

49. Jolly G, Smith R (2012) Volcanic Science Advisory Groups in New Zealand: an example from the recent eruption of Tongariro. In: Proceedings of Cities on Volcanoes, 19–23 November 2012, Colima, Mexico [http://www.citiesonvolcanoes7.com/vistaprevia2.php?idab=481], accessed 9th December 2014

50. Keough SM, Shanahan KJ (2008) Scenario planning: toward a more complete model for practice. Adv Dev Hum Resour 10:166-178 doi:10.1177/1523422307313311

51. (1998) Sources of Power: How people make decisions, 2nd printi. The MIT Press, Cambridge, Massachusetts; London, UK.

52. Klein G (2008) Naturalistic decision making. Hum Factors J Hum Factors Ergon Soc 50:456-460 doi:10.1518/001872008X288385

53. Kowalski-Trakofler KM, Vaught C, Scharf T (2003) Judgment and decision making under stress: an overview for emergency managers. Int J Emerg Manag 1:278-289 doi:10.1504/IJEM.2003.003297

54. Kozlowski SWJ (1998) Training and Developing Adaptive Teams. In: Cannon-Bowers JA, Salas E (eds) Making decisions under stress: Implications for individual and team training, American Psychological Association, Washington, D.C., USA. pp 115-153

55. Krauss W, Schafer MS, von Staunch H (2012) Introduction: post-normal climate science. Nat Cult 7(2):121-132

56. Lester P, Vogelgesang G (2012) Swift trust in Ad Hoc military organizations. In: Laurence J, Michael M (eds) The Oxford Handbook of Military Psychology, Oxford University Press, New York. pp 176-186

57. Lewis S (2012) Black Swan or Blind Spot? The Duality of Extreme Events. RISKworld: the newsletter of risktec solutions limited. 21:2. Accessed from http://www.risktec.co.uk/media/170409/riskworld%2021%20hr.pdf, 8th December 2014

58. Lindsay J, Marzocchi W, Jolly G, Constantinescu R, Selva J, Sandri L (2010) Towards real-time eruption forecasting in the Auckland Volcanic Field: application of BET_EF during the New Zealand National Disaster Exercise "Ruaumoko". Bull Volcanol 72:185-204, doi:10.1007/s00445-009-0311-9

59. Lipshitz R, Klein G, Orasanu J, Salas E (2001) Focus Article: Taking stock of naturalistic decision making. J Behav Decis Mak 14:331-352, doi:10.1002/bdm.381

60. Marks MA, Sabella MJ, Burke CS, Zaccaro SJ (2002) The impact of cross-training on team effectiveness. J Appl Psychol 87:3-13 doi:10.1037//0021-9010.87.1.3

61. Martin R (1992) Exercise Nga Puia Post-Exercise Report. Bay of Plenty Regional Council, Tauranga, NZ.

62. Martin L, Flin R, Skriver J (1997) Emergency decision making - A wider decision framework? In: Flin R, Salas E, Strub M, Martin L (eds) Decision making under stress emerging themes and applications, Ashgate Publishing Limited, Aldershot, England. pp 280-290

63. (2008) Exercise Ruaumoko '08 Final Exercise Report. Ministry of Civil Defence and Emergency Management, Wellington, NZ.

64. (2009) CDEM Exercises Director's Guideline for Civil Defence Emergency. Ministry of Civil Defence and Emergency Management, Wellington, NZ.

65. (2009) The Guide to the National Civil Defence Emergency Management Plan, 2006; revised June 2009. Ministry of Civil Defence and Emergency Management, Wellington, NZ.

66. (2010) Tsunami Advisory and Warning Plan. Management. Ministry of Civil Defence and Emergency Management, Wellington, NZ.

67. MCDEM website (2014a) National CDEM exercise programme 2010 [http://www.civildefence.govt.nz/cdem-sector/exercises/cdem-exercise-calendar/national-cdem-exercise-programme-2010/], accessed 6 October 2014

68. MCDEM website (2014b) National CDEM Exercise Programme [http://www.civildefence.govt.nz/cdem-sector/exercises/national-cdem-exercise-programme/], accessed 6 October 2014

69. MCDEM website (2014c) National CDEM Exercise Calendar [http://www.civildefence.govt.nz/cdem-sector/exercises/cdem-exercise-calendar], accessed 6 October 2014

70. McDowell S (2008) Exercise Ruaumoko: Evaluation Report. Auckland Civil Defence Emergency Management Group, Auckland, NZ.

71. Meyerson D, Weick KE, Kramer RM (1996) Swift trust and temporary groups. In: Kramer RE, Tyler TR (eds) Trust in organisations: frontiers of theory and research, Sage Publications, Inc, Thousand Oaks, CA, US. pp 166-195

72. Moats JB, Chermack TJ, Dooley LM (2008) Using Scenarios to Develop Crisis Managers: Applications of Scenario Planning and Scenario-Based Training. Adv Dev Hum Resour 10:397-424 doi:10.1177/1523422308316456

73. Morell VW, Sharp PC, Crandall SJ (2002) Creating student awareness to improve cultural competence: creating the critical incident. Med Teach 24:532-534 doi:10.1080/0142159021000 012577

74. Nairn IA (2010) Okataina Volcanic Centre Geology. [http://www.gns.cri.nz/Home/Learning/Science-Topics/Volcanoes/New-Zealand-Volcanoes/Volcano-Geology-and-Hazards/Okataina-Volcanic-Centre-Geology], accessed 23 April 2014

75. Natural Hazards Research Platform (2009) [http://www.naturalhazards.org.nz/], accessed 23 April 2014

76. Omodei MM, McLennan J, Elliott GC, Wearing AJ, Clancy JM (2005) "More is better?": A bias toward overuse of resources in Naturalistic decision making settings. In: Montgomery H, Lipshitz R, Brehmer B (eds) How professionals make decisions, Lawrence Erlbaum Associates, Mahwah, NJ. pp 29-41

77. Orasanu J, Connolly T (1993) The reinvention of decision making. In: Klein GA, Orasanu J, Calderwood R, Zsambok CE (eds) Decision making in action: models and methods, Ablex, Norwood, NJ. pp 3-20

78. Owen C, Campus SB, Brooks B, Chapman J, Paton D, Hossain L (2013) Developing a research framework for complex multi-team coordination in emergency management. Int J Emerg Manag 9:1-17

79. Pascual R, Henderson S (1997) Evidence of Naturalistic Decision Making in Military Command and Control. In: Zsambok CE, Klein G (eds) Naturalistic decision making, Lawrence Erlbaum Associates, Mahwah, NJ. pp 217-226

80. Paton D (1996) Training disaster workers: promoting wellbeing and operational effectiveness. Disaster Prev Manag 5(5):11-18

81. Paton D (2014) Disaster Management for Community Workers. Ministry of Health and Welfare, Taipei, Taiwan.

82. Paton D, Auld T (2006) Resilience in Emergency Management: Managing the flood. In: Paton D, Johnston D (eds) Disaster Resilience: An integrated approach, Charles C Thomas Publisher, Ltd, Springfield, Illinois, USA. pp 267-287

83. Paton D, Flin R (1999) Disaster stress: an emergency management perspective. Disaster Prev Manag 8:261-267 doi:10.1108/09653569910283897

84. Paton D, Jackson D (2002) Developing disaster management capability: an assessment centre approach. Disaster Prev Manag 11:115-122 doi:10.1108/09653560210426795

85. Paton D, Johnston DM, Houghton BF (1998) Organisational response to a volcanic eruption. Disaster Prev Manag 7:5-13 doi:10.1108/09653569810206226

86. Paton D, Johnston DM, Houghton BF, Smith LM (1998) Managing the effects of a volcanic eruption. Psychological perspectives of Integrated Emergency Management. J. Am Soc Prof Emerg Plan 5:59-69

87. Paton D, Johnston DM, Houghton BF, Flin R, Ronan K, Scott B (1999) Managing natural hazard consequences: planning for informaiton management and decision making. J. Am Soc Prof Emerg Plan 6:37-47

88. Paton D, Smith L, Violanti J (2000) Disaster response: risk, vulnerability and resilience. Disaster Prev Manag 9:173-180 doi:10.1108/09653560010335068

89. Pliske RM, McCloskey MJ, Klein G (2001) Decision skills training: facilitating learning from experience. In: Salas E, Klein G (eds) Linking expertise and naturalistic decision making, Lawrence Erlbaum Associates, Mahwah, NJ. pp 37-53

90. Pollock C, Paton D, Smith D, Violanti J (2003) Team Resilience. In: Paton D, Violanti J, Smith L (eds) Promoting capabilities to manage posttraumatic stress: perspectives on resilience, Charles C. Thomas, Springfield, Ill. pp 74-88

91. Potter SH, Scott BJ, Jolly GE (2012) Caldera Unrest Management Sourcebook. GNS Science, Lower Hutt, NZ.

92. Quarantelli EL (1997) Ten criteria for evaluating the management of community disasters. Disasters 21:39-56

93. Robert LP, Denis AR, Hung YTC (2009) Individual Swift Trust and Knowledge-Based Trust in Face-to-Face and Virtual Team Members. J Manage Inf Syst 26(2):241-279

94. Rogalski J, Samurcay R (1993) Analysing communication in complex distributed decision making. Ergonomics 36:1329-1342

95. Rosenthal U, 't Hart P (1989) Managing Terrorism: The south Moluccan Hostage Takings. In: Rosenthal U, Charles MT, 't Hart P (eds) Coping with Crises: the management of disasters, riots and terrorism, Charles C Thomas, Springfield, Illinois, USA. pp 340-366

96. Saarinen TF, Sell J (1985) Warning and Response to the Mt St Helens Eruption. State University of New York Press, Albany, NY.

97. Saaty TL (2008) Decision making with the analytic hierarchy process. Int J Serv Sci 1:83-98

98. Salas E, Stout RJ, Cannon-Bowers JA (1994) The role of shared mental models in developing shared situational awareness. In: Gilson RD, Garland DJ, Koonce JM (eds) Situational awreness in complex systems: proceedings of a Cahfa Conference, Embry-Riddle Aeronautical University Press, Daytona Beach, FL. pp 298-304

99. Salas E, Cannon-Bowers JA, Blickensderfer E (1997) Enhancing reciprocity between training theory and practice: Principles, guidelines, and specifications. In: Improving training effectiveness in work organisations. Erlbaum, Hillsdale, NJ. pp 291-322

100. Salas E, Cannon-Bowers JA, Johnston JH (1997) How can you turn a team of experts into an expert team? Emerging training strategies. In: Zsambok CE, Klein G (eds) Naturalistic decision making, Lawrence Erlbaum Associates, Mahwah, NJ. pp 359-370

101. Sarna P (2002) Managing the spike: The command perspective in critical incidents. In: Flin R, Arbuthnot K (eds) Incident command tales from the hot seat, Ashgate Publishing Limited, Aldershot, England. pp 32-57

102. Schaafstal AM, Johnston JH, Oser RL (2001) Training teams for emergency management. Comput Human Behav 17:615-626 doi:10.1016/S0747-5632(01)00026-7

103. Schwartz P (1996) The art of the long view: Planning for the future in an uncertain world. Currency Doubleday. Reprint Ed, New York, NY.

104. Siegrist M, Cvetkovich G (2000) Perception of hazards: The role of social trust and knowledge. Risk Anal 20:713-719

105. Sinclair H, Doyle EE, Johnston DM, Paton D (2012) Assessing emergency management training and exercises. Disaster Prev Manag 21:507-521 doi:10.1108/09653561211256198

106. Sinclair H, Doyle EEH, Johnston DM, Paton D (2012) Decision-making training in local government emergency management. Int J Emerg Serv 1:159-174 doi:10.1108/20470891211275939

107. Smith R (2009) Research, Science and Emergency Management: Partnering for Resilience. Tephra, Community Resilience: Research, Planning and Civil Defence Emergency Management. Ministry of Civil Defence & Emergency Management, Wellington, New Zealand.

108. Taleb NN (2007) The Black Swan: The Impact of the Highly Improbably. Random House Publishing Group, New York.

109. Thomson K, von Solms R, Louw L (2006) Cultivating an organizational information security culture. Comput Fraud Secur 2006:7-11 doi:10.1016/S1361-3723(06)70430-4

110. Tickell C (1990) Human effects of climate change: Excerpts from a lecture given to the Society on 26 March 1990. Geogr J 156:325-329

111. (2013) Taranaki Civil Defence Emergency Management Group [Joint Committee], Meeting Notes 26 November 2013. Taranaki Regional Council, Stratford, NZ.

112. UNESCO website (2014a) Exercise Pacific Wave 13, International Tsunami Information Centre [http://itic.ioc-unesco.org/index.php?option=com_content&view=category&id=2106&Itemid=2422], accessed 23 April 2014

113. UNESCO website (2014b) International Exercises, International Tsunami Information Centre [http://itic.ioc-unesco.org/index.php?option=com_content&view=category&layout=blog&id=1439&Itemid=1439], accessed 23 April 2014

114. UNESCO website (2014c) Exercise Pacific Wave 08, International Tsunami Information Centre [http://itic.ioc-unesco.org/index.php?option=com_content&view=category&layout=blog&id=1395&Itemid=1395], accessed 6 October 2014

115. Volpe CE, Cannon-Bowers J, Salas E, Spector PE (1996) The Impact of Cross-Training on Team Functioning: An Empirical Investigation. Hum Factors J Hum Factors Ergon Soc 38:87-100 doi:10.1518/001872096778940741

116. Waikato Regional Council (2014) The Caldera Advisory Group [http://www.waikatoregion.govt.nz/Services/Regional-services/Regional-hazards-and-emergency-management/Caldera-Advisory-Group-CAG/], accessed 23 April 2014

117. White C, Plotnick L, Addams-Moring R, Turoff M, Hiltz SR (2008) Leveraging A Wiki To Enhance Virtual Collaboration. The Emergency Domain. In Proceedings of the 41st Annual Hawaii International Conference on System Sciences (HICSS 2008). 1-10

118. Whittlesey R, Buckner N (1993) NOVA In the Path of a Killer Volcano. NOVA, PBS, USA.

119. Wilson I, Ralston W (2006) The scenario planning handbook: Developing strategies in uncertain times. South-Western Educational, Belmont, CA.

120. Wilson KA, Salas E, Priest HA, Andrews D (2007) Errors in the heat of battle: taking a closer look at shared cognition breakdowns through teamwork. Hum Factors 49:243-256

121. (2014) Reporting on the Seminar – Risk Interpretation and Action (RIA): Decision Making Under Conditions of Uncertainty. Austr J Disaster Trauma Stud 18(1):27-38

122. Zsambok C (1997) Naturalistic Decision Making: Where are we now? In: Zsambok CE, Klein G (eds) Naturalistic decision making, Lawrence Erlbaum Associates, Mahwah, NJ. pp 3-16

A Survey on Communication Technologies and Requirements for Internet of Electric Vehicles

Islam Safak Bayram[1] and Ioannis Papapanagiotou[2]

[1]Qatar Environment and Energy Research Institute, Qatar Foundation, Doha, Qatar

[2]Computer and Information Technology, Purdue University, West Lafayette, IN 47907, US

ABSTRACT

Electric vehicles (EVs) are becoming a more attractive transportation option, as they offer great cost savings, decrease foreign oil dependency, and reduce carbon emissions. However, varying temporal and spatial demand patterns of EVs threatens power grid operations and its physical components. Thus, the ability of the power grid to handle the potential extra load has become a major factor in the mainstream success. In order for this integration to occur seamlessly, the power grid and the consumers need to be coordinated in harmony. In this paper, we address the critical challenges introduced by the penetration of EVs, systematically categorize the proposed frameworks for demand management, and the role of information and communication technologies in the solution process. We provide a comprehensive survey on the communication requirements, the standards and the candidate technologies towards the Internet of electric vehicles (IoEV). This survey summarizes the current state of research efforts in electric vehicle demand management and aims to shed light on the continued studies.

REVIEW

Introduction

As the dependence on a single energy source (crude oil) exposes economies to unstable global oil market and increases environmental concerns, there has been a growing interest to push electric vehicles into mainstream acceptance. The motivation for the electrification of transportation is multifaceted; electricity can be generated through diverse and domestic resources, electricity prices have been relatively stable in the last two decades, and electric miles are cheaper and cleaner [1,2]. Therefore, internet of electric vehicles are expected to achieve a sizable market portion in the next decade. In fact, the study in [3] estimates that there will be around 50 million grid-enabled vehicles by year 2040.

Accordingly, there is a pressing need in the deployment of charging networks to accommodate the projected demand. For instance, [4] presents that there is an attempt to build a statewide charging station network in California. Similarly, Estonia is building the Europe's largest fast-charging station network with 200 nodes [5]. The number of EV charging stations is expected exceed four million in Europe and 11 million in the Globe by year 2020 [6].

However, as the power grid is becoming more congested due to the introduction of EVs, managing and controlling of corresponding demand should be carefully aligned with the available resources. Even though, the long term solution involves the upgrade of the power grid components, by considering the potential cost of such investments, the practical solution for the near term would be to develop intelligent control and scheduling techniques to aid the power grid operations. The realization of such frameworks requires appropriate communication architectures that will enable reliable interaction between the grid and the EV drivers to optimally control power flow under varying network conditions.

A handful of surveys have attempted to discuss general smart grid communication requirements, standards, and protocols for household demand management [7-10]. However, the case for the EVs is unique; electric vehicles can be mobile and a typical EV demand is large and, in fact, it can be more than the daily energy consumption of two households [11]. More importantly, the sustainability of the power grid operations is essential for human life. Therefore, careful attention is required to shed more light on the complex problems associated with electrictrification of transportation. Nonetheless, to the best of the authors' knowledge, this is the first study that focuses on the electric vehicle network communications for smart grid applications and, more specifically, to the IoEV challenges. Hence, in this work we

- Comprehensively address the unique challenges introduced by the EV penetration specifically for each power grid components and identify opportunities to improve the grid operations and system reliability;
- Systematically classify the mathematical frameworks for optimal control and management of EV demand; and
- Survey the communication requirements, standards, and candidate technologies that could serve the IoEVs and smart grid applications.

The structure of this paper is as follows. In Section 2, we present the current status of the U.S. power grid, the projected EV roll-out, potential negative impacts on power generation, transmission network, and distribution grid. Next, in Section 3, we categorize the literature on control and coordination frameworks according to the objective function, employed model, and the scale of the problem. In Section 4, we classify published standards and communication technologies respect to each smart grid application. In Section 5, we discuss the communication requirements and performance metrics for the IoEV network communications.

INTERNET OF ELECTRIC VEHICLES AND THE CURRENT POWER GRID

Internet of Electric Vehicles

Over the last few years, the automotive industry has introduced a variety of new electric vehicle models that have drastically expanded the customer choices [11]. The main drivers that shape the EV adoption include the size of the battery packs (usually varies between 16 to 56 kWh) and the duration to recharge the vehicle. The battery pack determines the all-electric range of the vehicle and, hence, it is an important criterion to beat the range anxiety. On the other hand, the charging duration depends on the employed charger technology, and it becomes a critical element in order to be competitive against the gas-powered counterparts. For instance, during a charging period of 30 min, level II single, and three-phase, and DC fast charge can enable a Nissan Leaf model (Nissan Motor Co., Ltd., Yokohama, Japan) to drive 5.5, 11, and 83.4 miles, respectively [12]. The charging standards may vary from country to country, and we present an overview of the different charger standards in Table 1. Moreover, the popularity of each charging type will greatly be determined by the housing demographics [13]. For instance, in the early EV adopter cities, a substantial portion of the population lives in multi-unit dwellings and EVs in these locations will likely use public fast charging facilities. Furthermore, several studies are conducted by different organizations to forecast the EV penetration rates. Depending on the assumptions made, prediction results may diverge, but nevertheless there is a consensus that EVs will represent a sizable portion in the next decades. In Table 2, the projected EV roll-out is presented. In the rest of this section, we present the current status of the power grid, potential impacts of EV demand, and opportunities offered optimal management of EVs.

Table 1: Electric vehicle charger technologies [14]

Type	Connection	Power (kW)	Max current
Europe	1-Phase AC	3.7	16 to 20
Europe	1 or 3 Phase AC	3.7 to 22	16 to 32
Europe	3-Phase AC	>22	>32
Europe	DC Fast	>22	>3.225
USA	AC Level-1	1.44	12
USA	AC Level-2	7.7	32
USA	DC Fast	240	400

Table 2: Electric vehicle penetration scenarios (approximate in millions) by different organizations

Year	US EIA - USA	NRC (probable) - USA	IEA world
2015	1 million	1.5 million	1.1 million
2020	2.3 million	3 million	6.9 million
2025	3.2 million	7 million	17.7 million
2030	4 million	14 million	33.3 million

US EIA: United States energy information administration [2, 15]; NRC: National Research Council [2, 15]; IEA: International Energy Agency [3].

Bayram and Papapanagiotou

Bayram and Papapanagiotou *EURASIP Journal on Wireless Communications and Networking* 2014 2014:223 doi: 10.1186/1687-1499-2014-223

Power Generation and Electricity Prices

Current Status

According to the US National Academy of Engineering, the power grid is 'the supreme engineering achievement of the twentieth century'. Currently, there are close to 3,200 utility companies serving more than 143 million customers in the United States. In order to serve the increasing customer demand, the required power supply is generated through diverse resources, including coal, nuclear, hydro, natural gas, and lately renewable sources, such as wind and solar [16]. Depending on the efficiency and the unit generation cost, power generation can be roughly divided into base load, intermediate load, and peak hour load. Factors that affect to dispatch a specific generation asset include variable operation and maintenance (O&M) costs, flexibility (fast vs. slow start generators), environmental 'head-room', and the distance to load and transmission. To meet the base load demand, utilities employ large scale ($\geq$400 MW) and low cost generation assets (e.g., nuclear, hydro, coal). Moreover, base load generation is characterized by high load factor (the percentage of hours that a power plant runs at full capacity) [17]. For intermediate load generation (the difference between expected customer demand and base load generation), power plants with lower load factors (typically around 50%) such as combined cycle combustion turbine fueled by natural gas etc. are employed [2,18]. Finally, utilities may need to employ additional generation assets to accommodate customer demand during peak hours. For this purpose, fast start, high cost, and usually environmentally unfriendly assets are employed. They are characterized by low load factor (5% to 10%), that leads to decreased utilization and hence and increased ratio of peak to average demand. Consequently, the use of such assets gradually increases the average kWh electricity price. A real-world scenario is illustrated in Figure 1a.

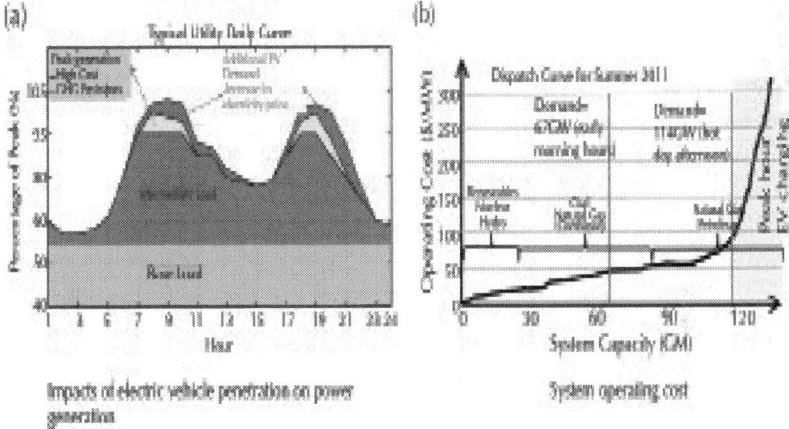

Figure 1: Impacts of EV charging on power generation and system operating cost. (a) Impacts of electric vehicle penetration on power generation. (b) System operating cost.

Impact of the EV Penetration

There are a handful of studies investigating the impact of electric vehicle charging on power generation [19-23]. According to [21], plug-in hybrid electric vehicles (assuming all vehicles are PHEV20 with a battery pack of 7.2 kWh) can increase the total load by 2.7% and the peak load by 2.5% in Colorado. On the other hand, battery sizes of pure EVs range from 16 to 52 kWh, which means actual impacts will be more severe. Similarly, [24] presents that if 5 % of the EV population charge at the same time, there will be a 5 GW increase in total power demand by year 2018 in VACAR region (Virginia - North Carolina - South Carolina). Overall, uncontrolled EV charging will decrease the utilization of low cost generation assets, increase the peak to average load ratio, and increase the power generation cost. Potential impacts of EV demand on the cost of the power grid is presented in Figure 1b.

Opportunities

The aforementioned effects can be mitigated with the deployment of necessary smart grid communication technologies which enable

EV users to take advantage of low prices during off-peak hours. In such applications, known as valley filling, grid operators encourage customers to postpone their EV charging to low power demand periods aiming to increase the overall power grid efficiency. There are many opportunities to use valley filling applications. The US power grid uses its maximum generation only around 5% of the time [25]. If optimal valley filling programs are employed, almost 73% of the vehicles in the US can be substituted by EVs [26]. Such an approach mandates EVs to be charged during the night when the aggregated power demand is low. For instance, the authors in [27] propose an EV charging framework for valley-filling applications in New York State with varying EV market penetrations of 5% to 40%. They show that the intelligent scheduling of EV chargings at off-peak hours increases the utilization of low cost generations, hence lowers the wholesale energy cost. In a similar study, authors of [23] argue that the savings gained due to intelligent charging of EVs could be reflected in charging tariffs and it promotes EV ownership. Furthermore, the work presented in [28] proposes a valley-filling algorithm and models the customer to grid interaction via pricing demand signals.

Transmission Network

Current Status

The transmission network ties the bulk power generation with the end users via high voltage lines. The US national grid includes three distinct geographic interconnections, namely the Eastern Interconnection, the Western Interconnection, and the Electric Reliability Council of Texas. The transmission network is composed of 170,000 miles of transmission lines rated at 200 (kV) and above, delivering the power generated at 5,000 (approximately) power plants [2]. Over the last two decades, the transmission network acts as an open highway which connects wholesale electricity markets to with end users. The primary goal of the network operators, on the other hand, is to make sure that transmission lines operate efficiently and reliably as it delivers the minimum cost generation to end users.

Impact of the EV Penetration

According to a study conducted by the US Department of Energy [29], in the Western Interconnection network alone, one third of the lines experienced congestion at least once during the year of the study, and 17% of the lines are congested at least 10% of the times. This study also shows that the situation is even more severe in the Eastern Interconnection, as the infrastructure is older and the network is not designed for long distance delivery of power.

On the other hand, the growth in EV load along with the deployment of new generators requires a capacity expansion in the transmission network. However, due to economic and political reasons, the required investments may not be realized in the short term. Past experiences show that new transmission projects can cost up to billions of dollars and may be stalled if the cost allocation and the recovery of investments are not properly planned. To that end, uncontrolled EV demand will allow transmission bottlenecks to emerge. These bottlenecks will increase electricity costs and the risks of blackouts.

Opportunities

The introduction of bidirectional chargers enables electric vehicles to transfer energy back to the grid (V2G) or to other electric vehicles (V2V) [30]. The utilization of such ancillary services can aid the transmission operations, mainly by reducing the congestion during peak hours. For example, group of vehicles can sell back part of their stored energy to other EVs who are in urgent need. This way, energy trading via V2V will eliminate the need to draw power from bulk power plants and hence the associated power losses in transmission will be minimized. For instance, studies in [31, 32] present mathematical framework to model the interaction of energy trading in a V2V scenario, where the groups of EVs determine the amount of energy to exchange and negotiate on unit price.

Moreover, EVs can transport their stored energy from one location to another which can support the grid via V2G applications. For example, [33] provides a transmission network based on the capability Internet of vehicles to transfer energy to the regions of high energy consumption. This way, the required upgrades will be deferred and occur gradually over time.

Distribution Network

Current Status

The distribution network is the final portion of the power grid which interfaces with the consumers. It is responsible for reducing the high voltage carried by transmission lines to appropriate levels for end users with the use of transformers typically rated between 2 to 40 kV. Over the last decade, the distribution network has been running up against its operating limits. In the US, national grid almost 7% of the electrical energy is lost (mostly in the form of heat) between generation units and end users and distribution network is mostly responsible for this. The distribution system is the most interruption-prone component of the power grid. According to [2], more than three-fourths of service interruptions originates in the distribution level.

Impact of the EV Penetration

If charged at parking lots or customer premises, the distribution grid is the part where most electric vehicles will be attached to. Uncontrolled EV charging could stress the distribution grid and cause system failures such as transformer and line overloading deteriorate power quality (e.g., large voltage deviations, harmonics, etc.). Considering the fact that EV penetration is going to be geographically clustered, negative impacts will be more severe in certain regions [2, 34, and 35]. For instance, the US distribution grid is designed to meet three to five houses [36] per transformer. Since charging of one EV doubles the daily load of a typical house, further challenges will be faced by the additional load introduced by EVs. A very typical scenario is illustrated in Figure 2 where five houses are served by a 37.5-kVA transformer. If just two level-2 chargers are used concurrently, local transformer is going to be overloaded. The frequent occurrence of such events will increase power loses and voltage deviations, and decreases transformer lifetime (high loading leads to high operating temperature) [3, 37, 38]. In [35], the authors presented a comprehensive study on the impacts of variety of EV charging scenarios on the required transformer upgrades and transformer efficiency.

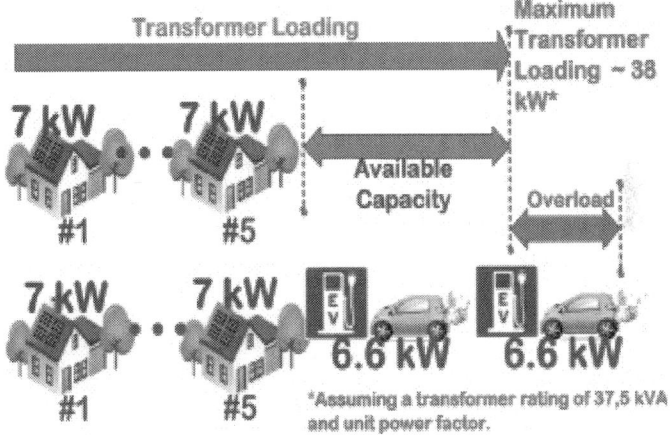

Figure 2: Potential distribution network overloading [39].

Opportunities

Intelligent control mechanisms (presented in the next section) can mitigate the aforementioned effects. Such frameworks requires both parties (EVs and the grid) to communicate. According to [40], controlling EV charging can reduce the number of congested (overloaded) network components which need to be replaced, hence eliminate the need for costly upgrades. It is further shown that controlling EV charging can reduce the cost of energy losses by 20% when compared to uncontrolled charging. In addition, EVs can be seen as distributed-energy storage mediums which are very essential for ancillary smart grid applications like integration of renewable energy resources and frequency regulation applications [41]. We provide a summary of the negative impacts of uncontrolled EV charging in Figure 3.

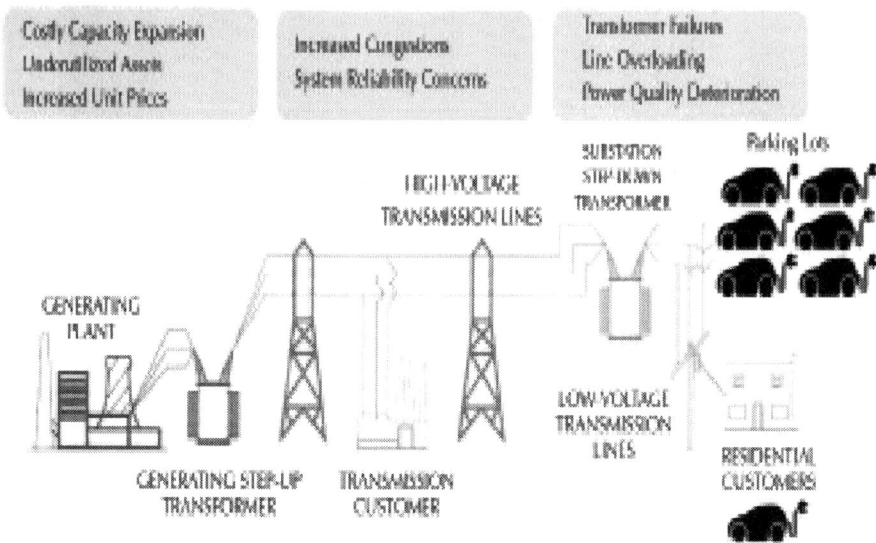

Figure 3: Impacts of uncontrolled EV charging.

DEMAND MANAGEMENT FOR THE INTERNET OF ELECTRIC VEHICLES

In order to mitigate the negative impacts of EV demand, there has been a growing interest in developing coordination strategies. At the heart of such frameworks lies information and communication technologies to support, control, and manage energy transfer between vehicles and the power grid that varies both in time and space, known as the Internet of EVs. In this section, we provide a comprehensive overview on the related literature. We classify the demand management techniques with respect to the objective of the optimization problem, scale of the problem, and the employed mathematical techniques. We present an overview of the literature in Figure 4 and the benefits of demand side management of EVs is summarized in Figure 5.

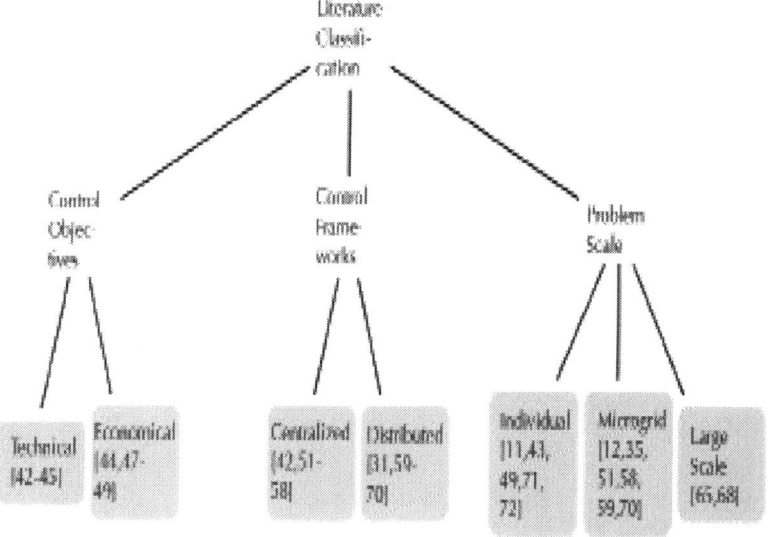

Figure 4: Literature classification of demand management techniques for Io-EVs.

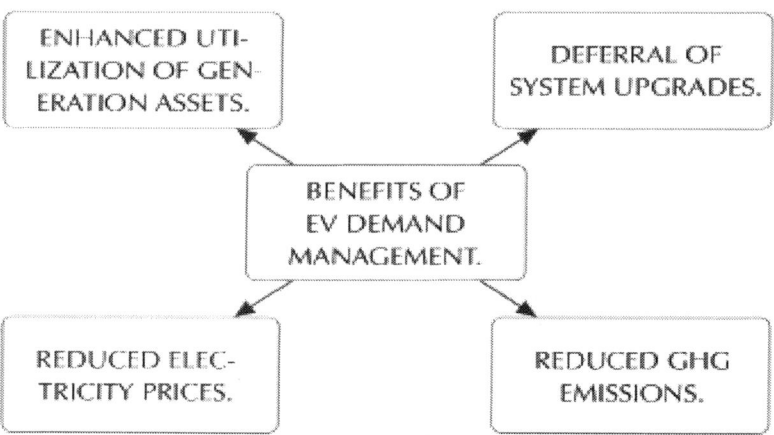

Figure 5: Benefits of electric vehicle management and control.

Control Objectives

Technical Objectives

The technical control objectives are usually related to the operating limits of the physical power grid assets. The most common objective functions are the minimization of energy losses, controlling voltage deviations, reducing peak-to-average load ratio, smoothing the consumer demand, and supporting renewable energy generation [42-45]. For vehicle-to-grid and vehicle-to-vehicle applications, the technical objectives include battery degradations and aging, thermal stability, etc. [46].

Economical Objectives

The objective functions fall into this category are usually linked to energy market participants: consumers, producers, retailers, etc. The main objectives include minimization of electricity generation and consumption costs. In this case, the objective functions are usually modeled with utility functions and the goal is to develop a charging tariff such that the total cost of charging is minimized compared to uncontrolled case [44, 47-49].

It is noteworthy that both of the objective functions are actually reflected in electricity prices. Hence, in some cases, technical objectives are coupled to economical objectives. Nodal pricing can be a good example [50], where the technical aspects (distance of generators, congestion of transmission lines, etc.) are translated into cost functions and the optimal pricing is solved with a more holistic approach.

Control Frameworks

The aforementioned control objectives are used in the mathematical frameworks to manage the EV demand. The applied control techniques depends on the employed charger technology. As given in Table 1, level I and level II charging typically takes a few hours, hence for these types, it is assumed that EVs are located in the customer's premises or at large parking lots. The majority of the literature considers EVs as 'smart' loads as the carving current can be adjusted in order to maximize the control objectives given above.

On the other hand, for the fast charging case, the EVs are assumed to be mobile and due to short service duration, the common control techniques include admission control at individual stations and customer routing/assignment in a network of charging stations. Overall, for both cases, the related literature can be divided into two categories: centralized and distributed controls.

Centralized Control

Centralized control employs a central authority (dispatcher) who up to a large extent controls and mandates EV charging rate, start time, etc. System level decisions, such as the desired state of charge, charging intervals, etc., are taken to finish all jobs by a certain deadline (e.g., by 7 am). Main advantages of centralized control include higher utilization of power grid resources and real-time monitoring of operation conditions across the network. On the other hand, to enable such functionalities, an advanced communication network is needed. Studies presented in [42, 51-56] are examples of centralized scheduling. These studies differ by the assumptions they make; interruptible vs. uninterruptible load, constant vs. varying charging rate, and preemptive vs. non-preemptive jobs. Management of EV fleets (e.g., school buses, postal service vehicles, etc.) can be a good example for centralized control. In this case, fleet owners can draw contracts with the utility operators and receive discounts. In return, utility can orchestrate EV demand according to network conditions to minimize his operating cost. Moreover, authors of [57] propose a deadline scheduling policy with admission control. They compare their algorithm with classic earliest deadline first and first come first serve scheduling. Similarly, the authors of [58] uses an admission control algorithm called Threshold Admission with Greedy Scheduling. In addition, their model incorporates renewable energy resources to charge electric vehicles.

Distributed Control

Decentralized control allows customers to choose their individual charging pattern. Decisions can be based on the price of the electricity or time of the day. This method eliminates the need of third party controller (dispatcher) and complex monitoring techniques. Since decisions are taken individually, game theoretic models are extensively employed. The works presented in [31,59,60] use Stackelberg game to model interactions of system operator (*leader*), who sets the prices and have the first move advantage, with individual EVs (*followers*) who respond to price changes by adjusting their demand. Another popular method is the Nash equilibrium, in which optimal pricing is achieved through maximization of individual utility functions [61, 62]. Other employed models include mean field games, potential games, and network routing games [61-70]. In addition to scheduling of night time charging, there is an interest in large scale charging of group stationary EVs (park and charge). For instance, [71] uses swarm optimization to allocate power to EVs in a parking lot. Authors of [72] propose a combined pricing-scheduling quadratic integer programming model to determine optimal prices and schedules to manage EV demand in large scale parking lots.

Scale of the Problem

The scale of the control framework can vary from individual level to entire transmission voltage level. We classify the scale of the problem into three categories.

- *EV scale:* This level of scale considers coordination of individual EVs according to the available information at the customer premises. Economical goals such as cost minimization and load profile smoothing are usually chosen [43, 49, 59, and 73].

- *Microgrid scale:* This level of problem considers groups of vehicles connected to LV/HV feeders. Typical examples include university campuses, parking lot (malls, airports, etc.), and microgrids. The control and coordination studies at this level include [12, 34, 58, and 71].

- *Transmission scale:* At this scale transmission, system operators and wholesale energy markets operate. Corollary, the control techniques applied considers thousands of EVs located in large geographical regions. The primary goal of this scale is to develop pricing policies to achieve optimal valley-filling during night time [62, 69].

AVAILABLE COMMUNICATION STANDARDS AND TECHNOLOGIES

The IoEV is based on the information and communication infrastructure to support the control and manage the energy transfer between vehicles and the power grid. In order to support such frameworks, we survey the related technologies and standards and the interdependency diagram which is presented in Figure 6. As this is a new area, some of the standards are either published or under development. We classify the communication standards and technologies into three groups: (1) the first group includes the technologies that are responsible for home charging applications and the message exchange between the EV and the charging equipment; (2) the second group includes the technologies for the mobile EV communication; and (3) the third group includes the standards for 'inter-control center' communication.

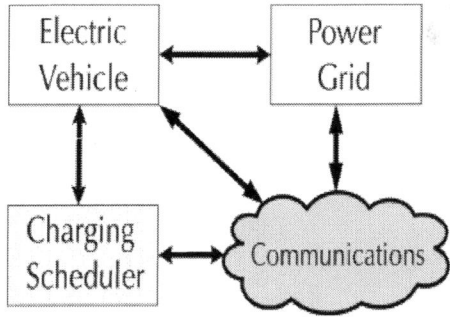

Figure 6: Interdependency of communications and EV demand management.

Communication Needs at Customer Premises

EV-electric Vehicle Supply Equipment

The communication at customer premises takes place in several places. First, group contains the standards and technologies between electric vehicle and electric vehicle supply equipment (EVSE) that is required for energy transfer monitoring and management, billing information, and authorization. The standardization is required for fast adoption of EVs and proper functioning of electric vehicle network components. The Society of Automotive Engineers (SAE) have defined the communication standards when an EV is being charged. We described these standards below [74, 75].

- *SAE J2293:* This standard covers the functionalities and architectures required for EV energy transfer system.

- *SAE J2836/1* and *J2847/1:* Define use cases and requirements for communications between EVs and the power grid, primarily for energy transfer. The central focus is on grid-optimized energy transfer for EVs to guarantee that drivers have enough energy while minimizing the reducing the stress on the grid.

- *SAE J2836/2* and *J2847/2:* Define the uses cases and requirements for the communications between electric vehicles and off-board DC charger.

- *SAE J2836/3* and *J2847/3:* Identify use cases and additional messages energy (DC) transfer from grid to electric vehicle. Also supports requirements for grid-to-vehicle energy transfer.

- *SAE J2931:* Defines digital communications requirements between EV and off-board device. SAE J2931/1 covers power line communications for EVs.

- *SAE J2931/2:* Defines the requirements for physical layer communications with in-band signaling between EV and EVSE.

In Figure 7, an overview of SAE communication standards is presented. For instance, J2836/1 use cases for utility programs may include time of use program, real-time pricing program, or critical peak pricing program [76]. Moreover, the International Electrotechnical Commission (IEC) is developing several standards under development for DC fast charging option. IEC 61851-23 presents the requirements for gird connections and communication architecture for fast charging. IEC 61851-24 defines the digital communications between EV and EVSE.

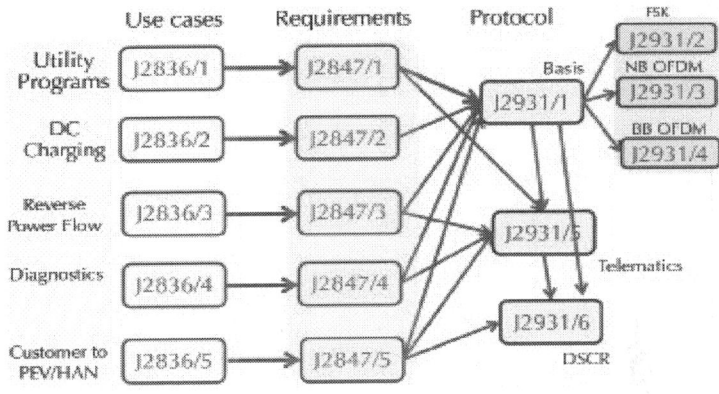

Figure 7: SAE communication standards.

Energy Management Unit to Power Grid

Visualization of energy consumption clearly helps customers to understand the cost of their energy usage. However, optimal decisions can only be taken by automated management systems [77, 78]. Energy management units (EMU) enables customers to power grid interaction; customers can monitor, control, and optimize their energy consumption. Even though energy management systems have been in the market for a few decades, the widespread adoption has gained pace with the recent advances in smart grid. [77] presents recent advances in EMUs.

EVSE will connect to EMU via home area network (HAN). The most popular technologies for HAN are Zigbee [79, 80], 802.11-based wireless local area network (WLAN), and femtocells. Zigbee offers required coverage (30 to 40 m), data rate (256 Kbps), low power usage, and deployment cost. In fact, it has a considerable market share in utility world [7,8]. The ubiquity of 802.11-capable devices makes WLAN a strong candidate for HAN. The details of WLAN technology is given in the next section. A comprehensive summary is presented in Table 3.

Table 3: Summary of candidate wireless technologies for IoEVs

	Latency	Throughput	Security	Scalability
WiFi				
IEEE 802.11a	L	H	M	M
IEEE 802.11b	L	M	M	L
IEEE 802.11g	L	H	M	L
IEEE 802.11n	L	H	M	M
3G				
UMTS/HSPDA	M	M	H	H
EVDO	M	M	L	H
4G				
LTE/HSPA+	L	H	H	H
IEEE 802.16e	L	H	H	H

Wireless mesh network can be implemented with WiFi nodes. Low (L): latency (< 250 ms), throughput (< 500 Kbps), scalability (< 100 nodes/backhaul node). Medium (M): latency (250 ms to 1 s), throughput (500 to 1,500 Kbps), and scalability (< 100 to 1,000 nodes/backhaul node). High (H): latency (> 1 s), throughput (> 1,500 Kbps), scalability (> 1,000 nodes/backhaul node).

Bayram and Papapanagiotou

Bayram and Papapanagiotou *EURASIP Journal on Wireless Communications and Networking* 2014 **2014**:223 doi: 10.1186/1687-1499-2014-223

Femtocells are usually employed as access points of cellular networks. This technology uses customer's broadband, DSL, etc. to connect to the wireless carrier's core network. This way, femtocells offer required indoor coverage and capacity for smart grid applications. Communication technologies with a special focus on security for home area networks is presented in [81].

For residential charging, the communications between EMU and the power grid is supported by the existing advanced metering infrastructure (AMI) network [82]. There are several candidates for this purpose.

Power line communications (PLC): PLC is a strong candidate for EMU to grid interaction. The main motivation for PLC is that already existing grid infrastructure reaches every EMU that wants to charge an EV. There are three different types of PLC technologies which are classified by the used frequency band and data rate. Broadband PLC uses 1.8 to 250 MHz frequency band and physical data rate varies between a few megabits to hundreds of megabits. Narrowband PLC operates in the 3 to 500 kHz band and provides lower data rates. Third type of PLC communications is ultra-narrow band technology, which is also the oldest type of all three. It only provides data rate around hundred bits per second [83].

Several millions of PLC-based communications have already been deployed globally [84]. Moreover, for EV to EVSE communications, PLC supports an apparent physical association that cannot be achieved by its wireless alternatives. Another distinctive advantage is that the cost of PLC deployment is relatively low when compared to other wireline options and can be comparable to wireless technologies.

However, there are several disadvantages for PLC. First, the communication medium is harsh and noisy. Second, transformers cause high attenuation which limits the range of the communication. Repeaters can be employed to overcome this problem, but additional cost should be taken into account beforehand. The final disadvantage is that regulations in some countries limits the use of PLC. For instance, PLC is not allowed for indoor environments in Japan [85].

White-space networking: The long term assignment of wireless spectrum to parties like digital TV broadcasters has created inefficient use of ISM band. Fatemieh, 2010 [86] proposes to use TV white spaces to meet communication requirements between users and the grid. IEEE 802.22 is the wireless regional area network (WRAN) standard that uses white spaces in the spectrum. The use of this technology offers the following benefits. It allows high data rates in a cost-effective way. White space networking has deep penetration and long range transmission capabilities, which would eliminate the need for complex designs (for EMU to data aggregation units). Also, high coverage can easily be achieved using white spaces. IEEE 802.11af, also referred to as 'White-Fi' and 'Super Wi-Fi' is a recent proposal that allow WLAN operation in TV white space spectrum in the VHF and UHF bands [87,88]. It uses cognitive radio technology to transmit on unused TV channels, with the standard taking measures to limit interference for primary users, such as analog TV, digital TV, and wireless microphones.

However, white-space networking is challenging. Available white spaces must be detected and interferences with the incumbents should be avoided. The underlying network should be able to run for varying bandwidths. Also, there are issues related to operation and management of the network [85, 86].

Wired infrastructure: Another option might be to build a wired infrastructure. Dedicated communication links give utilities full control over the network and reduce the reliance on the communication infrastructures operated by third parties. However, building such wired infrastructures is very costly. On the other hand, if the two-way communications is going to be a part of the power grid for the next century, it might be logical to build such an infrastructure gradually over time.

Customer's broadband: One school of thought suggests to use commodity broadband technologies, e.g., digital subscriber lines (DSL) or cable. The capital expenditures (CAPEX) for this case are lower, as the main communication infrastructure has already been deployed. Moreover, commodity broadband technologies uses Internet protocol (IP), so it can be easily connected to other ubiquitous IP-based communication networks. In a recent deployment, a DSL network was used as an underlying communication technology in Boulder, Colorado [89]. Nonetheless, there are several handicaps. The number of broadband connections is lower than the number of power meters. This is especially the case in developing countries. Moreover, the down times in some deployments is unacceptable for critical smart grid applications.

Other technologies: Mesh networks [85] have been proposed as alternative communication technology for AMI networks. Mesh networks tend to use different forms of wireless networks, i.e., IEEE 802.11, 3G/4G/5G, and mesh type of radio configuration. This choice is subject to technical, strategic, and even legal constraints. We present a detailed overview of such technologies in the next sections. In Table 4, we present an overview of candidate technologies and network technologies such as 3G/GSM, 4G/LTE (via smart apps such as [90,91]).

Table 4: Candidate communication technologies customer-to-grid interaction for garage charging

Technology	Pros	Cons
Power line communications	Every EV owner has an access. Easy penetration	Indoor applications are not allowed in every country. Regulatory and technical issues
White-space networking	High penetration and coverage	Require technologies to operate at varying bandwidths
Utility-owned wired infrastructure	Full control over the network. No need for interoperability among various standards	Very high cost and cost of ownership is not clear
Fixed broadband	Low cost (customers already have it)	Level of broadband deployment can be problematic
Wireless cellular networks	Easy adoption with already existing structure	Coverage is limited in developing countries
WiFi mesh network	Low cost, unlicensed frequency band	May require complex designs

Bayram and Papapanagiotou

Bayram and Papapanagiotou *EURASIP Journal on Wireless Communications and Networking* 2014 **2014**:223 doi: 10.1186/1687-1499-2014-223

An overview of the communication technologies for garage charging is presented in Figure 8 and summarized in Tables 5 and 6. Note that the communication requirements for the EV to EVSE is in the orders of milliseconds, while EVSE to EMU communication can occur in the order of seconds. Finally, the EMU can communicate with the grid in the order of minutes (typically every 15 min). In the next section, we will provide a comprehensive overview of such communication requirements.

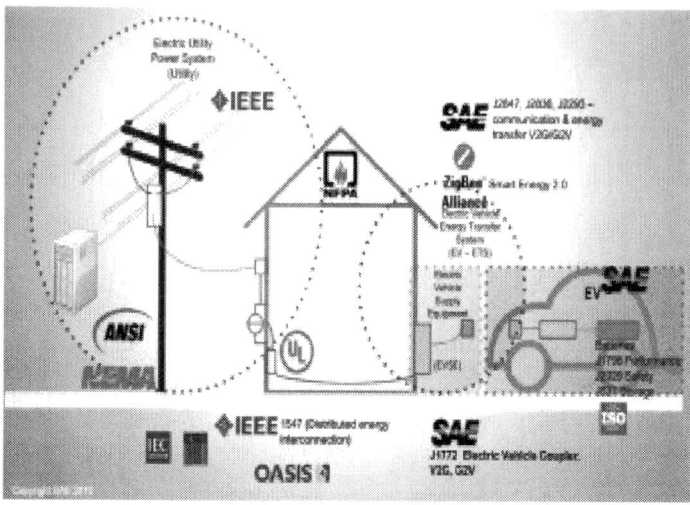

Figure 8: Overview of electric vehicle energy transfer standards (used with permission of SAE International [92]).

Table 5: Overview of communication standards for IoEVs

End users	Application	Name of standards and technologies
EV-EVSE	Energy transfer - garage charging	SAE J2293, SAE J2836/1, J2847/1, SAE J2836/2, J2847/2, SAE J2836/3, SAE J2847/3, SAE J2836/4, J2847/4, SAE J2931, IEC 61851-23, IEC 61851-24
EVSE - Energy Management Unit (EMU)	Home area network	Zigbee, 802.11, HomePlug
Customer (EMU) - grid	Garage charging, load shifting, valley filling, energy trading	PLC, 3G/4G/WiMAX/LTE/5G, WMN, TV white space, DSL, cable
Mobile EV - control center	Public charging	3G/4G/WiMAX/LTE/5G, WMN
Inter-control center	Public charging	IEC 60870-6/TASE.2

Bayram and Papapanagiotou

Bayram and Papapanagiotou *EURASIP Journal on Wireless Communications and Networking* 2014 **2014**:223 doi: 10.1186/1687-1499-2014-223

Table 6: Summary of findings: communications needs and requirements for IoEVs

Application	EV perspective		Grid perspective	
	Communication needs	Communication requirements	Communication needs	Communication requirements
Public charging	Locate and reserve charging station	High availability, service differentiation may be required	Load balancing among neighboring stations	QoS requirements increases with EV population
Residential charging	Respond to price updates to minimize charging cost	Part of AMI network (see[85])	Valley filling to better utilize power generation	Price updates sent every 15 min. Requirements for AMI hold
Energy trading via V2G	Sell part of stored energy to make profit or use stored energy during peak hours	High security and availability	Decrease the volume of storage medium needed by purchasing energy from EV fleets	The same as EV perspective

Bayram and Papapanagiotou

Bayram and Papapanagiotou *EURASIP Journal on Wireless Communications and Networking* 2014 2014:223 doi: 10.1186/1687-1499-2014-223

Mobile EV to Control Center Communications

Mobile EVs use public fast charging stations to fill up their batteries. Customer demand varies both spatially and temporally (e.g., downtown areas during rush hours) [37]. Also, the current status of the power grid limits grid operators to deploy the required number of charging stations. Hence, customer demand should be balanced among neighboring stations through the use of communication infrastructures. Thus, the ability to share data for mobile EVs becomes a necessity. In Figure 9, we present an overview of message exchange in electric vehicle networks.

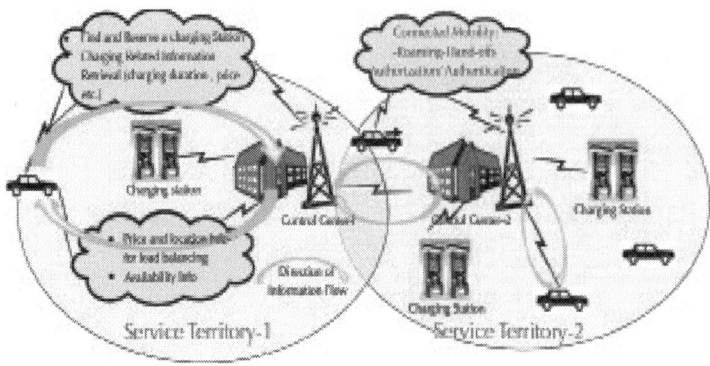

Figure 9: Internet of electric vehicles.

There are several wireless communication technologies that are projected to support 'electric mobility'. Two strong candidates are cellular network communications and wireless mesh networks.

Cellular Network Communications

For the short term, ubiquitous public cellular networks can provide required communication coverage in a cost-effective way. Moreover, cellular operators offer service solutions for smart grid applications. Power meter manufacturers embed communication modules to enable use of cellular communications. For garage charging and vehicle-to-grid applications data (e.g., power usage, price, etc.) are exchanged periodically (typically around every 10 to 15 min). Most cellular networks have sufficient capabilities to support the required communication medium. Further, cellular networks have the following advantages: (1) cellular communication technology is mature enough to meet smart grid needs; (2) since all cellular networks operate on licensed spectrum, there is no need to pay for unlicensed bands; and (3) cellular networks are scalable enough to connect huge number of EVs.

Worldwide interoperability for microwave access (WiMAX) is another candidate. WiMAX offers high capacity, wide coverage, low latency, low per-bit cost, and required quality of service capabilities. For example, garage charging applications generate small amount of traffic, but the projected number of connections is very high. For mobile

EVs, high data rate is needed to support location based applications. In most cases, in-vehicle application requires wide coverage, high throughput, and QoS support. WiMAX has required capabilities to handle the transmission of such data. In addition, mobile data service based on 4G long term evaluation (LTE) is becoming more popular as it can provides browsing experience comparable to wired connections. As of August 2013, there are more than 176 million LTE customers exist in the globe, and this number is expected to grow exponentially and exceed 1.3 billion by the end of 2018 [93]. Hence, 4G/LTE can provide a ubiquitous communication for EVs.

On the other hand, public charging applications require mobility support. As the mobile user moves faster, the supported data rate decreases. In Figure 10, we compare wireless communication networks according to mobility and throughput. 2.5G, 3G, 4G (WiMAX and LTE), and the upcoming 5G offer required connectivity for mobile EVs. IEEE P2030 standard [94] presents possible communication interfaces. The connection to central controller or telematics provider can be established by either equipment manufacturers OEMS or wide area communication.

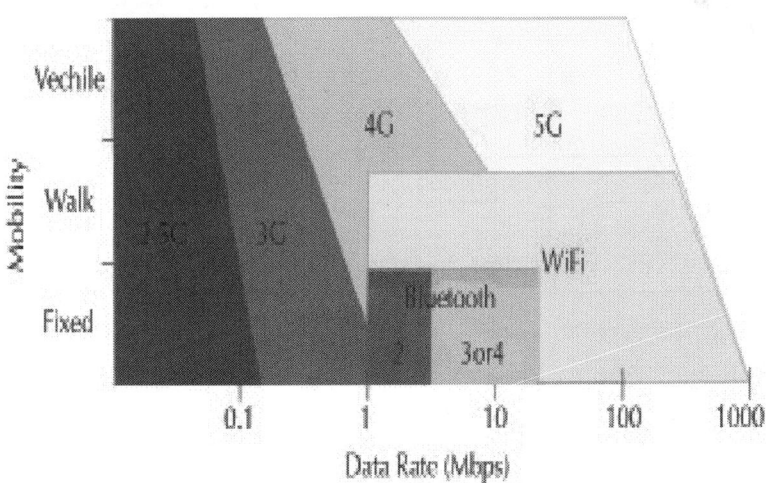

Figure 10: Data rate vs. mobility.

Wireless Mesh Networks

Wireless mesh networks (WMNs) are qualified to deliver required connectivity to EV drivers and the power grid. Moreover, their low cost, high scalability, self-healing, and self-organizing nature along with mobility support makes WMNs a very strong candidate. WMNs can provide high bandwidth and seamless handover capabilities at high speeds (almost the same quality as third generation technologies) [95]. Also, WMNs are compatible with other networks: they can be integrated with other existing networks (e.g., IEEE 802.15, IEEE 802.16, cellular networks, etc.). Further advantages of WMNs include its higher physical layer transmission rate than most cellular networks and coverage can be extended without using extra channel capacity.

Several companies already deployed WMNs for smart grid applications [96, 97]. As EV population continue to grow fast, the need for a dedicated communication infrastructure will become more important. Especially in urban environments, where 'xG' networks are overloaded or not deployed yet, WMNs will become even more important. In [97], a medium city is successfully deployed with wireless mesh networks to support required connectivity to electric vehicles.

On the other hand, WMNs have several disadvantages. In urban environments, network coverage can be affected by interference and fading. Available bandwidth can reduce in the case of possible loop problems [8]. In order to enjoy benefits of WMNs, research efforts are being shown to solve complexity of these networks.

Inter-control Center Communications

As shown in Figure 9, different regions are served by different service providers. Each control center monitors and controls registered customer demand at each charging facilities connected to him. Moreover, when a customer from another service territory requests service, control centers should be able to exchange information for authentication, billing, and location. Currently, all-electric range of most EVs is more than hundred miles [2]. This range enables drivers to go to different regions that are served by some other utility (e.g., Central Europe etc.). Hence, the communication network should be able handle possible hand-off situations.

At the present time, utilities employ IEC 60870-6/TASE.2 (International Electrotechnical Commission Tele-control Application Service Element) communication standard for information exchange between control centers, utilities, and power pools [8]. However, additional communication features may be needed to keep track of mobile users.

Further Communication Needs

Further, communication needs exist between EV and the charging equipment for the following periods: pre-charging, during the charging, and post-charging. In order to start the charging process, the EV and the charging equipment must be physically associated. Additional messages should be exchanged for identification and authorization purposes. During the charging, several parameters such as charging duration, direction of energy flow, available power and energy rate, vehicle status information (e.g., battery state of charge, usable battery energy, etc.) are needed to be exchanged between EV and EVSE. Precise measurement of transferred energy is also important for billing purposes [94].

COMMUNICATION REQUIREMENTS AND PERFORMANCE METRICS

The end-to-end communication requirements for EV network applications require highly available, reliable, and secure communications. Different applications, such as V2G, load shedding, etc., may have different communication requirements. The use cases for EV applications serve as a starting point for communication requirements. A detailed use case analysis is presented in [98, 99]. Each use case scenario defines the end-users (e.g., customer, utility, EV, etc.), their types (e.g., individual, organization etc.), content, size, and the frequency of the required message exchange. In this section, we discuss communication system requirements and associated performance metrics.

System Reliability and Availability

The successful management of EVs requires extensive use of reliable and (highly) available IoEV. The loss of availability is going to terminate the grid to customer interaction. During these isolation periods, customers will not be able to receive electricity prices, hence cannot optimally adjust and schedule their electricity usage. In fact, the cost of unavailability can be more severe. For instance, for garage charging scenarios, uncontrolled EV charging may lead to unwanted peaks and may overload some of the grid components, such as the distribution transformer.

Considering the aforementioned use cases, [100] explores the reliability requirements for home charging EV applications. The authors show that 11 different messages are used, and the minimum reliability requirement varies "between" 98.8% to 99.5%. This variety is attributed to some messages, such as vehicle identification number (VIN) information request, error messages related to EV charging rate, require high availability than other types.

The connectivity loss for mobile EVs is even more critical. Unavailability will refrain customers from locating and scheduling charging stations. Similarly, it may lead to suboptimal station selection both for customers (more expensive) and the grid operator (busy stations or long waiting lines may cause customer dissatisfaction) [101,102]. There are a handful of studies that quantify the cost of bad communication system performance. For instance, garage charging applications use AMI network. In a related study, [103] presents a generic AMI communication network and performs availability analysis for each component (e.g., home area network, 3G network, etc.). Moreover, it quantifies the cost of unavailability due to suboptimal power allocation.

There exist quite a few studies that present the performance evaluation of related wireless communication technology (e.g., UMTS etc.) [104-106]. A similar approach can be applied to mobile EV networks to quantify the cost of suboptimal charging station selections. On the other hand, redundancy design may help to improve system reliability. Employing redundant communication links between critical nodes such as data aggregation units to utility or between control centers. We present the overall system in Figure 11.

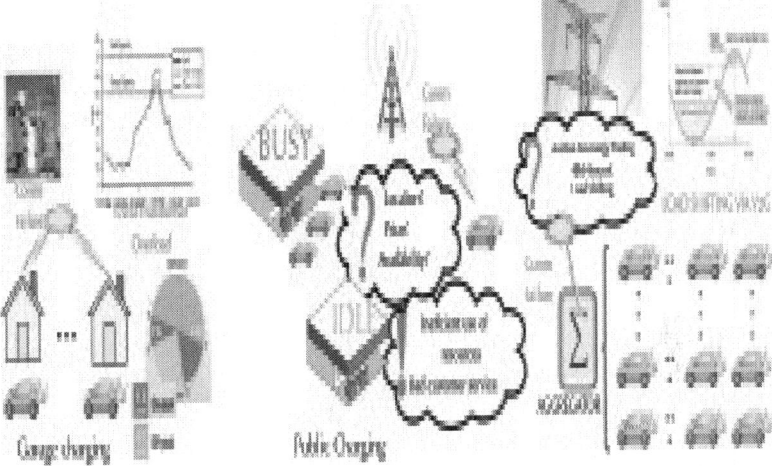

Figure 11: The negative effects of communication unavailability. Left panel: uncontrolled charging [2], middle panel: suboptimal charging station selection, and right panel: unable to support required storage medium for load shifting [107].

Quality-of-Service

The quality-of-service (QoS) needs are gradually increasing as the EVs gain widespread acceptance. Since centralized or decentralized control of EVs is done via price signals, degradation in communication system performance may cost. In [108], authors define QoS requirements for general smart grid communications using in terms of communication delays and outage probability.

The QoS requirements can be slightly different for mobile EVs and the grid operator. For instance, IEEE P2030 [94] states that an EV can afford to have a few seconds of latency to retrieve location, pricing, and availability information. However, in order to respond to the huge number of queries (approximate number depends on the EV penetration level) grid operator have to receive the information in a timely manner.

Even though today's mobile broadband technologies (e.g., 3G/HSPA/EV-DO etc.) promise high throughput and low latency communications, in some occasions, there can be a degradation in the user experience. This is attributed to the network capacity saturation in some areas. For instance, [109] shows that customer demand is going to exceed network capacity, for most metropolitan areas, in the next years. This will force time critical data transfer from EVs to compete with other bandwidth demanding applications such as video streaming and voice over IP.

On the other hand, the most recent mobile WiMAX/LTE technology can support necessary QoS requirements. More specifically, WiMAX offers four different QoS level, namely [110,111] (1) unsolicited grant service (UGS); (2) real-time polling service (rtPS); (3) non-real time polling service (nrtPS); and (4) best effort (BE). UGS can support low latency and low jitter and prioritize EV charging related data transfer. However, 4G technologies are not available everywhere and a limited but growing number of devices support 4G connectivity. Finally, some discussion is already undergoing about new 5G technologies [112].

In some areas, wireless mesh networks have been deployed using different versions of the IEEE 802.11 protocol. The cost of building such infrastructure is not expensive and does not require permission, since they function in the open 2.4 GHz or 5 GHz band. These networks can provide application access priority (starting from 802.11e and more recently with the 802.11ac), but they do not guarantee any strict QoS [113-116]. In addition, they have a limited range, which means that vehicles that want to communicate through them may be in a wireless blind spots.

Cyber-Physical Security

The power grid is vital to human life and with the integration of information systems, the power grid becomes a huge cyber-physical system. The grid's unique nature poses new series of security challenges. The components of the power grid are vulnerable to a variety of new cyber-security threats that could affect national security, public safety, and revenues.

There has been an increasing interest in smart grid security aspect [117-126]. In [120], the authors present cyber-physical security overview of smart grid communication infrastructure. Su, 2012 [119] presents security threats for electric vehicle networks. They conclude that electric vehicle networks have the following security requirements: (1) availability (discussed in the previous section); (2) confidentiality (prevent attackers to obtain private information); (3) integrity (block unauthorized users from changing the data); and (4) authenticity.

If the security of the EV network communication is not provided at a high level, an adversary can impact the EV network in various ways. A hacker can route customers to a specific charging station to create chaos for drivers. Similar to a home appliance, the garage charging is also programmed to fill up EV battery when price is low. An adversary can launch an attack to inject negative prices to increase power usage (of automated appliances), which may result in a peak or spike in electricity usage. Similarly, price modification can cause instabilities in V2G energy trading.

In [126], the authors present the security threats in physical layer of wireless communications for smart grid applications. Moreover, [125] defines the attack types for smart grid communication networks. They introduce three different kinds of smart grid attacks:

- *Data injection:* The type of attacks in this category falsify the meter measurements (e.g., garage charging) to mislead the power grid operator. The main purpose of this type of attack is to create revenue loss.

- *Vulnerability:* This type of attack is caused by the failure of a communication channel or a device. Information on the feedback channel can be unsynchronized due to erroneous communication links.

- *Intentional:* In this type, the attacker has the full knowledge of network topology. It can be carried out by targeting the node with the highest degree with a denial-of-service attack.

Several organizations including IEEE (1402-2000, IEEE Guide for Power Substation Physical and Electronic Security), North American Electrical Reliability Corporation - Critical Infrastructure Protection (NERC-CIP), National Infrastructure Protection Plan (NIPP), and National Institute of Standards and Technology (NIST) [118]. In the second volume of NISTIR 7628 [122], NIST documents a comprehensive overview of guidelines for smart grid cyber-security. This documents contains several use cases concerning the security issues with EV charging. In [124], the authors evaluated the effectiveness of NISTIR framework for an electric vehicle charging infrastructure case. They claim that NISTIR 7628 framework is not strong enough in device authentication and protecting the protecting the location privacy of mobile EVs.

Scalability

As the EV population is continuously going to increase for the next couple of decades, the underlying communication networks should be scalable enough to support required functionalities. Such scalability concerns can be alleviated by employing IP-based network designs. Considering the big smart grid picture on mind, it is very likely that that required communication networks will be based on IPv6. Moreover, IP-based solutions offer huge cost savings in deployment and maintenance [7].

Capacity

Since EV applications generate data traffic, the underlying communication networks should be have enough capacity to meet minimum communication requirements. For mobile EVs, the required capacity can be measured in bits-per-second. However, for residential charging applications, the communication capacity is more likely to be measured in the maximum number of advanced meters (or smart meter) that it can support at a time (since most messages types/lengths are standard).

In a related study [127], researchers analyze the capacity of a linear chain network topology for an AMI network. They also compute the required network capacity for different amounts of nodes, varying message lengths, and meter reading periods (e.g., every 10 or 15 min). They also extend their study for larger networks with different communication infrastructures.

On the other hand, capacity comes at the expense of cost. Capacity planning is a critical step as it includes trade-offs that could affect the success of EV applications. Initial deployments may seem easy and does not require high capacity networks, since EV population will be low. This will allow utilities to have a good head start with low installation cost. However, short term solutions are likely to fail to scale. Hence, the expected exponential growth in EV population may force utilities to replace the entire communication network.

Interoperability

The proper functioning of EV-related applications depends on different entities such as power system and communication system to work together. According to the US Independence and Security Act (2007), the NIST is appointed to be the main global coordination of such smart grid interoperability.

In its framework [128], NIST identifies the domains of the smart grid as: customers, markets, service providers, operations, bulk generation, transmission, and distribution. NIST's conceptual framework also provides the required information exchange between these domains. EV applications are unique in the sense that they bridge most of these domains. For instance, home charging deals with distribution network and the service provider, V2G deals with markets, and public fast charging is related to bulk transmission and customers.

IEEE P2030 Smart Grid Interoperability Series of Standards aims to establish an interoperability framework to develop IEEE-based standards on power system applications and control through the use of communication infrastructures. The first of this series IEEE Std 2030 (2011) presents communication and information networks interfaces for different domains of the smart grid. Moreover, this reference model presents the communication requirements for each interface (e.g., security, availability, latency etc.).

In addition, the IEEE P2030.1 Working Group [129] develops a draft guide for electric-sourced transportation infrastructures. Also, P2030 task force-3 defines communication requirements between devices in the smart grid. They are going to describe the network, transport, and session layers (from OSI reference model). Recently, IEEE has established a new technical advisory group (IEEE 802.24) which will work with multiple IEEE 802 working group standards of which are very essential for smart grid communications [130].

Measurement-based Studies

Previous paragraphs show that wide-area wireless communication technologies will be predominant role in EV network communications. On the other hand, since the number of mobile internet users has flourished, the user experience deviated significantly from theoretical results. Hence, there is a need for detailed measurement based studies to understand and predict the performance of the wireless technology and quantify the effects of performance degradation.

There are only a handful of measurement-based studies that focuses on the performance of the wireless network (WiFi, 3F (UMTS), EV-DO, and WiMAX) [131-133]. In [133], authors conducted a measurement study to evaluate the performance of the mobile Internet access with 3G (UMTS) and WiFi networks. The measurement was carried out in Seattle, San Francisco, and Amherst. Across all cities, the average availability of 3G and WiFi is 87% and 11%, respectively. The details of their findings is presented in Table 7. Then, they proposed a hybrid framework to improve the availability of 3G by augmenting it with WiFi.

Table 7: Availability performance of wide area wireless technologies [133]

	Amherst		Seattle		San Francisco
	Average	**Peak**	**Average**	**Peak**	**Average**
3G (UMTS)	90%	85.5%	82%	79%	89%
WiFi	12%	10%	10%	8.5%	11%

Bayram and Papapanagiotou

Bayram and Papapanagiotou *EURASIP Journal on Wireless Communications and Networking* 2014 **2014**:223 doi: 10.1186/1687-1499-2014-223

Similarly, [132] presents an architecture to improve end user experience by exploiting (i) channel diversity, (ii) wireless network service provider diversity, and (iii) technology diversity (UMTS, CDMA, etc.). Their results shows that the proposed *Mobile Access Router* architecture decreases the blackout periods considerably and increases average throughput. In addition, [131] shows the results of a city-wide mobile Internet experimentation results. The mobile nodes in their test bed employs both EV-DO and WiFi interfaces. Their focus is on measuring the signal latency and TCP throughput performance. Their results indicate that average latencies varies between 150 to 400 ms and mobile TCP throughput is around 752 Kbps.

CONCLUSIONS

In this paper, we provided a survey of the communication requirements and technologies for the Internet of electric vehicles. First, we presented the current status of the power grid. We specifically focused on the power generation and distribution networks. We identified the challenges introduced by the projected EV demand. Then, we showed that the EV demand may have disruptive effects in the current information and the IoEV infrastructures that are needed to support, control, and manage the energy transfer between vehicles and the power. Next, we grouped related smart grid applications and surveyed the communication requirements, standards, and candidate technologies for each group. We showed that in the absence of two-way communications, the proliferation of EVs will pose threats to the existing power grid and will not reach projected mainstream success.

In the future, we plan to expand our research in the following ways. The choice of communication technology and standards should consider the performance of the each candidate. It is also worth noting that the importance of performance evaluation will increase as the EVs gain widespread acceptance. For instance, if a central authority receives a few queries (location and pricing information for public charging stations) per minute, the cost of communication delays, unavailability, etc. will be negligible. On the other hand, as the query rate increases, underlying infrastructure should provide high availability and low latency. Thus, it is crucial to quantify the effects of the underlying communication technology.

REFERENCES

1. Electrification roadmap: Revolutionizing transportation and achieving energy security. Technical report (2009)

2. JG Kassakian, R Schmalensee, The future of the electric grid: an interdisciplinary MIT study. Technical report, Technical report, Massachusetts Institute of Technology (2011)

3. Technology roadmap: Electric and plug-in hybrid electric vehicles. Technical report, International Energy Agency (June 2011)

4. Tesla To Build National Electric Car Charging Network (http://www), .forbes.com/sites/toddwoody/2012/09/25/tesla-to-build-national-electric-car-charging-network

5. ABB Wins Tender for Europes Largest Electric Vehicle Fast-charging Network (http://www), abb.com/cawp/seitp202/d07e075541462e04c

6. Electric vehicles charging equipments. Technical report, Pike Research (2011)

7. Y Yan, Y Qian, H Sharif, D Tipper, A survey on smart grid communication infrastructures: Motivations, requirements and challenges. IEEE Commun. Surv. Tutor 15(1), 1–16 (2012)

8. V Gungor, D Sahin, T Kocak, S Ergut, C Buccella, C Cecati, G Hancke, A survey on smart grid potential applications and communication requirements. IEEE Trans. Ind. Inform 9(1) (2012)

9. W Su, H Eichi, W Zeng, MY Chow, A survey on the electrification of transportation in a smart grid environment. IEEE Trans. Ind. Inform 9(1), 1–10 (2012)

10. W Wang, Y Xu, M Khanna, A survey on the communication architectures in smart grid. Comput. Netw 55(15), 3604–3629 (2011).

11. IS Bayram, G Michailidis, M Devetsikiotis, B Parkhideh, Strategies for competing energy storage technologies for in dc fast charging stations. in *Proc*, ed. by . IEEE International Conference on Smart Grid Communications (Tainan City, Taiwan, 2012), pp. 1–6

12. IS Bayram, G Michailidis, M Devetsikiotis, Electric power resource provisioning for large scale public EV charging facilities. in *Proc*, ed. by . IEEE International Conference on Smart Grid Communications (Vancouver, Canada, 2013)

13. E Bloom, Global building stock database: Commectical and residential building floor space by country and building type (2011-2012). Technical report, Pike Research (2012)

14. MC Falvo, D Sbordone, M Devetsikiotis, Bayram I S, EV charging stations and modes: international standards. in *Proc*, ed. by . IEEE International Symposium on Power Electronics, Electrical Drives, Automation and Motion (Naples, Italy, 2014)

15. Transitions to Alternative Transportation Technologies–Plug-in Hybrid Electric Vehicles. The National Academies Press (http://

www, 2010),. nap.edu/openbook.php?record_id=12826

16. GT Heydt, R Ayyanar, KW Hedman, V Vittal, Electric power and energy engineering: the first century. Proc. IEEE 100(Special Centennial Issue), 1315–1328 (2012)

17. GW Arnold, Challenges and opportunities in smart grid: a position article. Proc. IEEE 99(6), 922–927 (2011)

18. B Wollenberg, A Wood, *Power generation, operation and control* (John Wiley&Sons, Inc, 1996)

19. M Scott, M Meyer, D Elliot, W Warwick, Impacts of plug-in hybrid vehicles on electric utilities and reginal US power grids. Technical report, Pasific Northwest National Laboratory, Palo Alto, CA (2007)

20. P Denholm, W Short, An evaluation of utility system impacts and benefits of optimally dispatched plug-in hybrid electric vehicles. Technical report, National Renewable Energy Laboratory (2006)

21. K Parks, P Denholm, T Markel, Costs and emissions associated with plug-in hybrid electric vehicle charging in the Excel energy colorado service territory. Technical report, National Renewable Energy Laboratory (2007)

22. SRW Letendre, M Cross, Plug-in hybrid vehicles the vermont grid: a scoping analysis. Technical report, University of Vermont Transportation Center (2008)

23. A Shortt, OM Malley, Quantifying the long-term impact of electric vehicles on the generation portfolio. IEEE Trans. Smart Grid 5(1), 71–83 (2014)

24. WH Hadley, Impact of plug-in hybrid vehicles on the electric grid. Technical report, Oak Ridge National Labs (October 2006)

25. A Ipakchi, F Albuyeh, Grid of the future. IEEE Power Energy Mag 7(2), 52–62 (2009)

26. W Shireen, S Patel, Plug-in hybrid electric vehicles in the smart grid environment. in *Proc*, ed. by . IEEE PES Transmission and Distribution Conference and Exposition (IEEE, 2010), pp. 1–4

27. K Valentine, WG Temple, and KM Zhang, Intelligent electric vehicle charging: rethinking the valley-fill. J. Power Sources 196(24), 10717–10726 (2011).

28. L Gan, U Topcu, S Low, Optimal decentralized protocol for electric vehicle charging. Proc. IEEE Conference on Decision and

Control and European Control Conference, 5798–5804 (2011)

29. S Abraham, National transmission grid study. Technical report (2002)

30. C Liu, KT Chau, D Wu, S Gao, Opportunities and challenges of vehicle-to-home, vehicle-to-vehicle, and vehicle-to-grid technologies. Proc. IEEE 101(11), 2409–2427 (2013)

31. W Tushar, W Saad, HV Poor, DB Smith, Economics of electric vehicle charging: a game theoretic approach. IEEE Trans. Smart Grid 3(4), 1767–1778 (2012)

32. Y Wang, W Saad, Z Han, HV Poor, T Basar, A game-theoretic approach to energy trading in the smart grid. IEEE Trans. Smart Grid 5(3), 1439–1450 (2014)

33. P Yi, T Zhu, B Jiang, B Wang, D Towsley, An energy transmission and distribution network using electric vehicles. Proc. IEEE International Conference on Communications, 3335–3339 (2012)

34. K Clement-Nyns, E Haesen, J Driesen, The impact of charging plug-in hybrid electric vehicles on a residential distribution grid. IEEE Trans. Power Syst 25(1), 371–380 (2010)

35. S Shao, M Pipattanasomporn, S Rahman, Challenges of PHEV penetration to the residential distribution network. Proc. IEEE Power Energy Society General Meeting, 1–8 (2009)

36. A Kwasinski, A Kwasinski, Signal processing in the electrification of vehicular transportation: techniques for electric and plug-in hybrid electric vehicles on the smart grid. IEEE Signal Process. Mag 29(5), 14–23 (2012)

37. MD Galus, MG Vayá, T Krause, G Andersson, The role of electric vehicles in smart grids.Wiley Interdiscip. Rev. Energy Environ 2(4), 384–400 (2013)

38. K Clement-Nyns, E Haesen, J Driesen, The impact of charging plug-in hybrid electric vehicles on a residential distribution grid. IEEE Trans. Power Syst 25(1), 371–380 (2010)

39. How the smart grid enables utilities to integrate electric vehicles (white paper). Silver Spring Networks (2012)

40. RA Verzijlbergh, MO Grond, Z Lukszo, JG Slootweg, MD Ilic, Network impacts and cost savings of controlled EV charging. IEEE Trans. Smart Grid 3(3), 1203–1212 (2012)

41. C Quinn, D Zimmerle, TH Bradley, The effect of communication architecture on the availability, reliability, and economics of plug-in hybrid electric vehicle-to-grid ancillary services. J. Power Sour 195(5), 1500–1509 (2010).

42. E Sortomme, MM Hindi, SDJ MacPherson, S Venkata, Coordinated charging of plug-in hybrid electric vehicles to minimize distribution system losses. IEEE Trans. Smart Grid 2(1), 198–205 (2011)

43. L Jian, GXu H Xue, X Zhu, D Zhao, ZY Shao, Regulated charging of plug-in hybrid electric vehicles for minimizing load variance in household smart microgrid. IEEE Trans. Ind. Electron 60(8), 3218–3226 (2013)

44. N Rotering, M Ilic, Optimal charge control of plug-in hybrid electric vehicles in deregulated electricity markets. IEEE Trans. Power Syst 26(3), 1021–1029 (2011)

45. J Wang, C Liu, D Ton, Y Zhou, J Kim, A Vyas, Impact of plug-in hybrid electric vehicles on power systems with demand response and wind power. Energy Policy 39(7), 4016–4021 (2011).

46. M Yilmaz, PT Krein, Review of the impact of vehicle-to-grid technologies on distribution systems and utility interfaces. IEEE Trans. Power Electron 28(12), 5673–5689 (2013)

47. X Xi, R Sioshansi, Using price-based signals to control plug-in electric vehicle fleet charging. IEEE Trans. Smart Grid 5, 1–15 (2014)

48. P Samadi, A-H Mohsenian-Rad, R Schober, VWS Wong, J Jatskevich, Optimal real-time pricing algorithm based on utility maximization for smart grid. In *Proc*, ed. by . IEEE International Conference on Smart Grid Communications (Washington D.C., USA, 2010), pp. 415–420

49. IS Bayram, M Abdallah, K Qaraqe, Providing QoS guarantees to multiple classes of EVs under deterministic grid power? Proc. IEEE International Energy Conference (2014)

50. L Chen, H Suzuki, T Wachi, Y Shimura, Components of nodal prices for electric power systems. IEEE Trans. Power Syst 17(1), 41–49 (2002).

51. IS Bayram, G Michailidis, M Devetsikiotis, F Granelli, Electric power allocation in a network of fast charging stations? IEEE J.

Selected Areas Commun 31(7), 1235–1246 (2013)

52. M Alizadeh, A Scaglione, RJ Thomas, from packet to power switching: digital direct load scheduling. IEEE J. Selected Areas Commun 30(6), 1027–1036 (2012)

53. AY Saber, GK Venayagamoorthy, Efficient utilization of renewable energy sources by gridable vehicles in cyber-physical energy systems. IEEE Syst. J 4(3), 285–294 (2010)

54. H Lu, G Pang, G Kesidis, IE, CS&E and EE Depts, Automated scheduling of deferrable PEV/PHEV load in the smart grid. Technical report. Technical Report CSE-12-004, Pennsylvania State University, CSE Dept (2012)

55. RA Waraich, 1 plug-in hybrid electric vehicles and smart grid: investigations based on a micro-simulation Transportation Res. Part C: Emerging Technol 28, 74–86 (2013)

56. K Clement, E Haesen, J Driesen, Coordinated charging of multiple plug-in hybrid electric vehicles in residential distribution grids. Proc. IEEE Power Systems Conference and Exposition, 1–7 (2009)

57. S Chen, Y Ji, L Tong, Large scale charging of electric vehicles. Proc. IEEE Power and Energy Society General Meeting, 1–9 (2012)

58. S Chen, Y Ji, L Tong, Deadline scheduling for large scale charging of electric vehicles with renewable energy. Proc. IEEE Sensor Array and Multichannel Signal Processing Workshop, 13–16 (2012)

59. I Bayram, G Michailidis, I Papapanagiotou, M Devetsikiotis, Decentralized control of electric vehicles in a network of fast charging stations. in *Proc*, ed. by . IEEE International Global Communication Conference (Atlanta, GA, 2013)

60. IS Bayram, G Michailidis, M Devetsikiotis, Unsplittable load balancing in a network of charging stations under QoS guarantees. IEEE Trans. Smart Grid 6, 1–11 (2014)

61. L Gan, U Topcu, S Low, Optimal decentralized protocol for electric vehicle charging. Proc. IEEE Conference on Decision and Control and European Control Conference, 5798–5804 (2011)

62. DS Callaway, IA Hiskens, Achieving controllability of electric loads. Proc. IEEE 99(1) (2011)

63. Z Ma, D Callaway, I Hiskens, Decentralized charging control for large populations of plug-in electric vehicles. *Decision and*

Control (CDC), 2010 49th IEEE Conference On (IEEE, 2010), pp. 206–212

64. N Rotering, M Ilic, Optimal charge control of plug-in hybrid electric vehicles in deregulated electricity markets. IEEE Trans. Power Syst 26(3), 1021–1029 (2011)

65. MD Galus, G Andersson, Demand management of grid connected plug-in hybrid electric vehicles (phev). Proc. IEEE Energy 2030 Conference, 1–8 (2008)

66. IS Bayram, M Ismail, M Abdallah, K Qaraqe, E Serpedin, A pricing-based load shifting framework for EV fast charging stations. IEEE International Conference on Smart Grid Communications (2014)

67. L Gan, U Topcu, SH Low, Stochastic distributed protocol for electric vehicle charging with discrete charging rate. Proc. IEEE Power and Energy Society General Meeting, 1–8 (2012)

68. L Gan, U Topcu, S Low, Optimal decentralized protocol for electric vehicle charging. Proc. IEEE Decision and Control and European Control Conference, 5798–5804 (2011)

69. Z Ma, DS Callaway, IA Hiskens, Decentralized charging control of large populations of plug-in electric vehicles. IEEE Trans. Control Syst. Technol 21(1), 67–78 (2013)

70. Z Fan, Distributed charging of PHEV in a smart grid. Proc. IEEE International Conference on Smart Grid Communications, 255–260 (2011)

71. W Su, M-Y Chow, Performance evaluation of a PHEV parking station using particle swarm optimization. Proc. IEEE Power and Energy Society General Meeting, 1–6 (2011)

72. A Deshpande, P Murali, Pricing long-term permits and scheduling of electric vehicle charging in parking lots with shared resources. *Control Conference (ECC), 2013 European* (IEEE, 2013), pp. 3584–3589

73. L Jian, H Xue, G Xu, X Zhu, D Zhao, ZY Shao, Regulated charging of plug-in hybrid electric vehicles for minimizing load variance in household smart microgrid. IEEE Trans. Ind. Electron 60(8), 3218–3226 (2013)

74. SAE Vehicle electrictification standards (http://www), . sae.org/smartgrid/ [Accessed: June 2014]

75. K Gowri, RG Pratt, FK Tuffner, MCW Kintner-Meyer, Vehicle to grid communication standards development, testing and validation: status report. Pacific Northwest National Laboratory, Technical report (2011)

76. T Bohn, H Chaudhry, Overview of SAE standards for plug-in electric vehicle. Proc. IEEE PES Innovative Smart Grid Technologies, 1–7 (2012)

77. S Aman, Y Simmhan, VK Prasanna, Energy management systems: state of the art and emerging trends. IEEE Commun. Mag 51(1), 114–119 (2013)

78. L Bartram, J Rodgers, K Muise, Chasing the negawatt: visualization for sustainable living. Proc. IEEE Comput. Graph. Appl. Conf 30(3), 8–14 (2010)

79. Zigbee Alliance (http://www), . zigbee.org/ [Accessed: Aug. 2014]

80. J Mu, A minimum physical distance delivery protocol based on zigbee in smart grid. EURASIP J. Wireless Commun. Netw 2014(1), 108 (2014).

81. E Bou-Harb, C Fachkha, M Pourzandi, M Debbabi, C Assi, Communication security for smart grid distribution networks. IEEE Commun. Mag 51(1), 42–49 (2013)

82. Standardization roadmap for electric vehicles. Technical report, American National Standards Institute (April 2012)

83. S Galli, A Scaglione, Z Wang, Power line communications and the smart grid. Proc. IEEE International Conference on Smart Grid Communications, 303–308 (2010)

84. S Galli, A Scaglione, Z Wang, For the grid and through the grid: the role of power line communications in the smart grid. Proc. IEEE 99(6), 998–1027 (2011)

85. P Kulkarni, S Gormus, Z Fan, B Motz, A mesh-radio-based solution for smart metering networks. IEEE Commun. Mag 50(7), 86–95 (2012)

86. O Fatemieh, R Chandra, CA Gunter, Low cost and secure smart meter communications using the tv white spaces. Proc. IEEE International Symposium on Resilient Control Systems, 37–42 (2010)

87. D Lekomtcev, R Maršálek, Comparison of 802.11af and 802.22 standards - physical layer and cognitive functionality. Elektro Revue 3(2), 12–18 (2012)

88. AB Flores, RE Guerra, EW Knightly, P Ecclesine, S Pandey, Ieee 802.11af: a standard for TV white space spectrum sharing. IEEE Commun. Mag 51(10), 92–100 (2013)

89. Smart Grid DSL and QWEST Team Up (http://gigaom),. com/cleantech/smart-grid-dsl-current-and-qwest-team-up/ [Accessed: Aug. 2014]

90. Plugshare- EV Charging Station Map (http://www), plugshare. com [Accessed: Jan. 2014]

91. Electric Vehicle Charging (http://www), chargepoint.com [Accessed: Oct. 2013]

92. J Pkrzywa, SAE ground vehicle standards smart grid (SAE Taipei), . Available: http://sae-taipei.org.tw/image/1283265726.pdf

93. B Bangerter, S Talwar, R Arefi, K Stewart, Networks and devices for the 5G era. IEEE Commun. Mag 52(2), 90–96 (2014)

94. IEEE draft guide for smart grid interoperability of energy technology and information technology operation with the electric power system (EPS), and end-use applications and loads. IEEE P2030/D5.0 February 2011, 1–126 (2011)

95. ML Sichitiu, Wireless mesh networks: opportunities and challenges. Proceedings of World Wireless Congress (2005)

96. Tropos Networks (gridcom). tropos.com [Accessed: Aug. 2014]

97. Volkswagen Research (http://www). wireless-wolfsburg.de [Accessed: July 2014]

98. M Burns, Interoperability knowledge base (http://collaborate), nist.gov/twiki-sggrid/bin/view/SmartGrid/InteroperabilityKnowledgeBase

99. Edison Smartconnect - Industry resource center: 2008-2009 smart grid use cases (http://www). sce.com/CustomerService/smartconnect/industry-resource-center/smartgrid-usecase.htm

100. E Hossain, Z Han, HV Poor, *Smart Grid Communications and Networking* (Cambridge Univ. Press, Cambridge, UK, 2012)

101. IS Bayram, G Michailidis, M Devetsikiotis, S Bhattacharya, A Chakrabortty, F Granelli, Local energy storage sizing in plug-

in hybrid electric vehicle charging stations under blocking probability constraints. Proc. IEEE International Conference on Smart Grid Communications, 78–83 (2011)

102. IS Bayram, G Michailidis, M Devetsikiotis, F Granelli, S Bhattacharya, Smart vehicles in the smart grid: challenges, trends, and application to the design of charging stations? In *Control and Optimization Methods for Electric Smart Grids*, ed. by Chakrabortty A, Ilic MD. Power Electronics and Power Systems, vol. 3 (Springer, 2012), pp. 133–145

103. D Niyato, P Wang, E Hossain, Reliability analysis and redundancy design of smart grid wireless communications system for demand side management. IEEE Wireless Commun.19(3), 38–46 (2012)

104. S Dharmaraja, V Jindal, U Varshney, Reliability and survivability analysis for UMTS networks: an analytical approach. IEEE Trans. Netw. Serv. Manag 5(3), 132–142 (2008)

105. AP Snow, U Varshney, AD Malloy, Reliability and survivability of wireless and mobile networks. Computer 33(7), 49–55 (2000).

106. A Bruce, Reliability analysis of electric utility SCADA systems. Proc. IEEE Power Industry Computer Applications, 200–205 (1997)

107. S-I Inage, Modeling load shifting using electric vehicles in a smart grid environment. Technical report, OECD Publishing (2010)

108. H Li, W Zhang, QoS routing in smart grid. Proc. IEEE Global Telecommunications Conference, 1–6 (2010)

109. Mobile broadband capacity constraints the need for optimization. Technical report, RYSAVY Research (2010)

110. I Papapanagiotou, D Toumpakaris, J Lee, M Devetsikiotis, A survey on next generation mobile WiMAX networks: objectives, features and technical challenges. IEEE Commun. Surv. Tutor 11(4), 3–18 (2009)

111. GS Paschos, I Papapanagiotou, CG Argyropoulos, SA Kotsopoulos, A heuristic strategy for ieee 802.16 WiMAX scheduler for quality of service. 45th Congress FITCE (2006)

112. C-X Wang, F Haider, X Gao, X-H You, Y Yang, D Yuan, H Aggoune, H Haas, S Fletcher, E Hepsaydir, Cellular architecture and key technologies for 5G wireless communication networks. IEEE Commun. Mag 52(2), 122–130 (2014)

113. I Papapanagiotou, GS Paschos, SA Kotsopoulos, M Devetsikiotis, Proc. IEEE Global Telecommunications Conference, 2530–2535

114. I Papapanagiotou, GS Paschos, M Devetsikiotis, A comparison performance analysis of QoS WLANS: approaches with enhanced features. Adv. Multimedia 2007(1), 1 (2007)

115. GS Paschos, I Papapanagiotou, SA Kotsopoulos, GK Karagiannidis, A new MAC protocol with pseudo-TDMA behavior for supporting quality of service in 802.11 wireless LANs. EURASIP J. Wirel. Commun. Netw 2006(3), 8–189 (2006)

116. I Papapanagiotou, JS Vardakas, GS Paschos, MD Logothetis, SA Kotsopoulos, Performance evaluation of IEEE 802.11e based on ON-OFF traffic model. in *Proceedings of the 3rd International Conference on Mobile Multimedia Communications*, ed. by . MobiMedia '07 (ICST, Brussels, Belgium, Belgium, 2007), pp. 17–1176 (http://dl, 2007),. acm.org/citation.cfm?id=1385289.1385310

117. Y Yan, Y Qian, H Sharif, D Tipper, A survey on cyber security for smart grid communications. IEEE Commun. Surv. Tutor 14(4), 998–1010 (2012)

118. J Liu, Y Xiao, S Li, W Liang, CLP Chen, Cyber security and privacy issues in smart grids. IEEE Commun. Surv. Tutor 14(4), 981–997 (2012)

119. H Su, M Qiu, H Wang, Secure wireless communication system for smart grid with rechargeable electric vehicles. IEEE Commun. Mag 50(8), 62–68 (2012)

120. Y Mo, TH-J Kim, K Brancik, D Dickinson, H Lee, A Perrig, B Sinopoli, Cyber 2013; physical security of a smart grid infrastructure. Proc. IEEE 100(1), 195–209 (2012)

121. Roadmap to achieve energy delivery systems cyber security. Technical report, The US Department of Energy (2011)

122. NISTIR 7628 Guidelines for smart grid cyber security strategy, Architecture, and High-Level Requirements Technical report, National Institute of Standards and Technology

123. M Qiu, H Su, M Chen, Z Ming, LT Yang, Balance of security strength and energy for a PMU monitoring system in smart grid. IEEE Commun. Mag 50(5), 142–149 (2012)

124. AC-F Chan, Z Zhou, on smart grid cybersecurity standardization: issues of designing with NISTIR 7628. IEEE Commun. Mag 51(1), 58–65 (2013)

125. P-Y Chen, S-M Cheng, K-C Chen, Smart attacks in smart grid communication networks. IEEE Commun. Mag 50(8), 24–29 (2012)

126. E-K Lee, M Gerla, SY Oh, Physical layer security in wireless smart grid. IEEE Commun. Mag50(8), 46–52 (2012)

127. B Karimi, V Namboodiri, Capacity analysis of a wireless backhaul for metering in the smart grid. Proc. IEEE Conference on Computer Communications Workshops, 61–66 (2012)

128. N Framework, Roadmap for smart grid interoperability standards. National Institute of Standards and Technology (2010)

129. IEEE P2030.1 Draft Guide for Electric-Sourced Transportation Infrastructure (http://grouper), . ieee.org/groups/scc21/2030.1/2030.1_index.html [Accessed: Aug. 2014]

130. IEEE 802.24 Smart Grid Technical Advisory Group (http://standards), . ieee.org/news/2012/802.24tag.html

131. J Ormont, J Walker, S Banerjee, A Sridharan, M Seshadri, S Machiraju, A city-wide vehicular infrastructure for wide-area wireless experimentation. *Proceedings of the Third ACM International Workshop on Wireless Network Testbeds, Experimental Evaluation and Characterization* (ACM, 2008), pp. 3–10

132. J Ott, D Kutscher, A disconnection-tolerant transport for drive-thru internet environments. Proc. IEEE Computer and Communications Societies. Proceedings, vol. 3, 1849–1862 (2005)

133. A Balasubramanian, R Mahajan, A Venkataramani, Augmenting mobile 3G using WiFi. Proceedings of the 8th International Conference on Mobile Systems, Applications, and Services, 209–222 (2010)

Chapter 5

The Potentials of and Barriers to the Utilization of Advanced Computer Systems in Remote Construction Projects: Case of the Kingdom of Saudi Arabia

Bhzad Sidawi and Abdulsalam Alsudairi

Architecture Department, College of Architecture and Planning, University of Dammam, 31451 Dammam, Kingdom of Saudi Arabia

ABSTRACT

Background

An extensive body of knowledge indicated the positive impact of the Advanced Computer based Management Systems (ACMS) on various aspects of project management, while highlighting barriers that hinder adoption, diffusion, and utilization of the ACMS by the construction industries around the world. Remote projects have their unique management problems and these are caused mainly by the remoteness of the project. Little research was undertaken concerning this issue, particularly in the Persian Gulf region, and it has highlighted few unique communications and management problems such as the loose control, lack of human resources, infrastructure and experience.

Methods

This research investigated the use of ACMS by large companies in the Eastern province, Kingdom of Saudi Arabia (KSA), and how it would help these companies sorting out a number of present projects' management problems. Subsequently, a field study i.e. a questionnaire survey and interviews was carried out.

Result

The field study revealed significant association between frequent management problems with little use of ACMS and the domination of use of traditional communications and management systems. This paper argues that the use of traditional systems and the traditional way of sorting out construction problems limit the applicability of ACMS.

Conclusions

The present researchers recommend the use of customized ACMS associated with the application of lean and sustainable management principals as these would help overcoming barriers and providing intelligent solution for the strategic, technical, and social issues of the remote construction sites.

BACKGROUND

Remote construction projects exist in many regions throughout the world such as the Sahara desert, Antarctic regions, the Arabian Peninsula desert, the Australian desert, the Empty Quarter etc. The dilemma in managing remote projects is highlighted by Deng et al. (2001), Kestle and London (2002, 2003), Kestle (2009), McAnulty and Baroudi (2010), and Thorpe (2000). These authors have pointed out that the remoteness thus the loose control is major cause of the management problems. They suggested possible causes such as the lack of human resources, infrastructure and experience of how to manage these remote projects. In the Kingdom of Saudi Arabia, remote construction sites have unique problems and these have been highlighted by Justanyah and Sidawi (2011, Sidawi 2010a,b, 2012a,b.

The literature review suggested that some of the project management and communications problems are caused by the use of inappropriate tools and systems for communication, coordination, and management. For example, Yang et al. (2007) suggested that intense need for project information and effective communications by the project team cannot be met by traditional communications and information management systems since these systems have shortcomings and are incapable of fulfilling project duties and objectives. One of these shortcomings is that traditional systems provide limited access to information, which is considered one of the key barriers to successful project management practices (Vadhavkar and Pena-Mora 2002; Pena-mora et al. 2009). Many of these project failures are caused by inadequate organization and management of the construction process (e.g., a weak coordination of processes and uncertainty about available information) (Wamelink et al. 2002).

These management problems can be sorted out by the use of ACMS such as mobile, Web-based Project Management Systems (WPMS), augmented reality and BIM (Alshawi and Ingirige 2003 and Stewart and Mohamed 2004, Nitithamyong and Skibniewski 2004, 2006, Charoenngam et al. 2004, Arayici and Aouad 2010, Arayici et al. 2012). These systems use wireless, satellite, Internet-based, or mobile tools and networks and it helps - to a certain degree - construction industry firms manage the increasing complexity of normal construction projects. They have also helped fulfill project objectives such as quality, scope, time, and cost. Sustainable practices and measures would also help in sustaining the quality of projects, eliminate waste, and minimize the cost (Arayici et al. 2012). Previous research however, highlighted the barriers that affect the adoption, diffusion, and full application of the technology (Alshawi and Ingirige 2003, Nuria 2005; Leskinen2006, 2008, Nitithamyong and Skibniewski 2007).

In the KSA, some large companies such as Aramco and Royal commission of Jubail (RCJ) have a number of remote construction projects. These projects are of different sizes i.e. medium, large to very large. They are in remote locations and some operate in undeveloped and environmentally sensitive regions. They are far from the supervision team office, the contractor's office, and major urban concentrations. During construction, all project parties experience countless difficulties and cumbersome management problems. These potential problems negatively affect project quality and cause substantial delays and increases in costs. This paper investigates the potential use of ACMS by large companies in the Eastern province in managing remote projects and barriers that hinder the potential use, and how ACMS would help these companies in sorting out project's problems and improving their management practices.

LITERATURE REVIEW

Remote Construction Sites Problems

The dilemma of managing remote projects is highlighted by Deng et al. (2001), who mentions that the extensive physical distance between project participants, sometimes extending over national boundaries, is the primary cause of delays in decision making. The project team has to not only tackle traditional management problems but those that specifically occur as a result of the remote locations of these often environmentally sensitive sites (Kestle 2009; Kestle and London 2002,2003). These sites are often far from logistic support and suffer a continuous shortage of materials and specialized labor (Kestle and

London 2002, 2003). Kestle (2009) investigated the management problems of remote projects and reports lack of project pre-planning, certainty, and/or clarity concerning project process integration. There were also misinterpretations and miscommunications of project results and needs issues. A centralized decision-making process and lack of delegated authority to field personnel often hindered progress and communications at critical emergency response and recovery stages. Kestle and London (2002) suggested a framework for the design management of remote sites. The framework emphasizes the following management functions: serving, controlling, organizing and economizing.

McAnulty and Baroudi (2010) conducted a survey of top and mid-tier construction contractors with experience in remote construction projects in Australia. They find that contractors experience difficulty attracting and retaining skilled workers; working in remote locations has a negative impact on an employee's family life. It is difficult to procure and access materials and equipment in remote areas and severe climatic factors in remote areas have a negative impact on productivity. There is lack of infrastructure and communications. The researchers suggest a number of possible solutions such as the need for appropriate material management systems and design cost information specifically for remote construction works. They recommended that unique types of costing issues should be included in the project's cost estimation at the pre-construction stages of project; these include: mobilization/demobilization, accommodation, inclement weather downtime, site allowances, delivery, and productivity.

Justanyah and Sidawi (2011, and Sidawi 2010a,b, 2012a,b examined the case of Saudi Electric Company (SEC). SEC has a number of remote projects and it experiences difficulties in running these projects. These problems can be summarized as the following:

- Procurement and risk Management: There is a frequent shortage of materials. The delivery of materials and equipment is constrained by road/highway regulations and bad conditions of some remote roads. Projects have much higher risk margins than ordinary projects. This is caused by the ad hoc approach and both sides i.e. the owner and the contractor do not accurately plan projects.

- Cost, time, scope, and quality management: There is serious delay in decision-making, loose control, and infrequent visits to the remote site result in wasted time, excessive costs, unfocused scope, and poor construction quality.

- Human Resources: SEC has a staff shortage so employees are incapable of doing all required site visits. There is a lack of security and shortage in skilled workers. The impact of the harsh working conditions has negative impact on the productivity of SEC's supervisors.

- Infrastructure and communications: Land ownership in some remote areas is not definite or known; legal disputes are likely to occur. There is a lack of or no infrastructure. Contractors and SEC supervisors still use traditional communications and management tools. Decisions are made autocratically and SEC's project managers are not able to control and coordinate integration of a project's aspects and the typical management style is non-standard, fragmented, and loose.

The Potentials and Barriers to the Use of ACMS

The Potentials of ACMS

Some of a remote project management's problems can be avoided if ACMS were used. So, it is important to know how far these systems would be capable in sorting remote construction problems. These systems include Web based Project Management Systems WPMS and mobile systems that feature mobile tools, personal digital assistants (PDA), wearable computers, wireless tools, four dimensional augmented reality (4D), virtual reality and other technologies. These systems possess the capability to improve communications between project team members and enable teams to share information and quickly solve problems (Charoenngam et al. 2004). They improve team members' ability to manage time and costs (ibid). Davidson and Moshini (1990) and Bowden (2005) stated that construction costs can be reduced by 25% through efficient transfer of information between the construction teams; that transfer can be achieved through ACMS. Ahuja et al. (2010) suggested that adoption of Information Communication Technology (ICT) enables effective communication between dispersed project team members but argues that strategic adoption of ICT. This enables them to grasp effectively the IT benefits (Ahuja et al. 2009).

Alshawi and Ingirige (2003) and Stewart and Mohamed (2004) identified the following benefits of using WPMS: productivity enhancement of communication between project participants, reduction in project delays, heightened awareness of project issues among all parties, and ease of access to and retrieval of project information. Other advantages include: avoiding delays due to the arrival of updated drawings and documents, reducing visits to sites and travelling time to meetings, avoiding drawing mistakes, reducing time and money spent on disputes, sharing and exchanging project information, automating repetitive routine processes, and eliminating paper reports. Thomas et al. (2003) discussed how WPMS - from the point of view of selecting contractors - helps project managers boost contractor performance and confidence by minimizing subjectivity and eliminating the potential for corrupt practices. This improves competitiveness through increased awareness of competitors' strengths and weaknesses and nurtures mutual trust in the exchange of sensitive information such as performance data. Nitithamyong and Skibniewski (2004, 2006) suggested that benefits of using WPMS can be categorized into four main areas. These categories include cost reduction and time saving, enhancement of communications and collaboration, improvement of productivity and partnership, and support of e-commerce and the customer. A number of researchers anticipate that WPMS will replace traditional project management methods (Becerik 2005; Zou and Roslan 2005) and these methods are drivers of WPMS adoption. Several aspects support this claim including increased competitive pressures, expectations of revenue growth, the ability to compete globally, and the desire to reengineer the business to respond to market challenges (Nitithamyong and Skibniewski 2006).

Leskinen (2006, 2008) argued that it is difficult to make direct assessment of which mobile systems would benefit the construction industry. The most important intangible benefits include improving customer service, gaining a competitive advantage, acquiring more timely management information, supporting core business functions, avoiding competitive disadvantages, improving management information, improving product quality, improving internal communication, implementing changes through innovation, improving external communication, and enhancing the jobs of employees. In recent years, the development of laser scanning and video and image-based 3D reconstruction system is enabling remote and virtual walk through on actual construction sites. These systems have the ability to minimize the travel times of supervisors and may increase the frequency of progress, quality and safety inspections by providing project supervisors with systems that are easily applicable (Golparvar-Fard et al. 2011, Jaselskis et al. 2011). The utilization of Building Information Modeling (BIM) by the project team would provide a more streamlined business process, associated project and site management methodologies including complete facilitation of construction knowledge during the full lifecycle of a building project (Arayici and Aouad 2010). Arayici et al. (2012) argued that BIM implementation serves as a useful alternative to addressing key construction sector issues, and offer solutions to these in order to increase productivity, efficiency, quality; reduce costs, lead times and duplications, via effective collaboration and communication of stakeholders in remote construction projects. These key findings, in terms of challenges, also lend support to the classification of the key challenges for construction project management in remote construction projects, such as human resources, cost, time, scope, and quality management; procurement and risk management, infrastructure and communication.

Justanyah and Sidawi (2011) and Sidawi 2010a,b, 2012a,b have studied the potentiality of Communications and Project Management Systems (CPMS) and whether it would help SEC in avoiding some construction problems, sorting out efficiently the site queries and improving the management of these projects. They pointed out that the CPMS would support SEC's management activities, particularly during the construction stages that witness long delays in providing feedback to the construction enquiries. CPMS can be used to provide fast feedback to the site personnel regarding the most problematic construction problems such as these related to the monitoring of the construction process, quality of work, procurement of materials and productivity levels.

The section above highlighted the problems and capabilities of ACMS regarding each of the project management aspects (see also Table 1). The previous research indicated that remote construction sites have unique problems that need unique and customized IT solutions. It also indicated that remote construction sites have some common problems such as these related to time, potential risk, pre-planning, cost, and logistic support despite that they are in different geographical locations.

Table 1: The management problems of remote construction projects and potential ACMS solutions

Management problems	Potential ACMS solutions
The extensive physical distance between project participants, sometimes extending over national boundaries, is the primary cause of delays in decision making (Deng et al., 2001)	ACMS can improve team members' ability to communicate and manage time and costs (Charoenngam et al.,2004)
Lack of project pre-planning, certainty, and/or clarity concerning project process integration (Kestle, 2009)	ACMS use would reduce cost and save time, enhance communications and collaboration, improve productivity and partnership, and support the e-commerce and the customer (Nitithamyong and Skibniewski, 2004, 2006)
These sites are often far from logistic support and suffer a continuous shortage of materials and specialized labor (Kestle and London, 2002, 2003). Contractors experience difficulty attracting and retaining skilled workers; and it is difficult to procure and access materials and equipment and site's conditions has a negative impact on productivity (McAnulty and Baroudi, 2010)	Construction costs can be reduced by 25% through efficient transfer of information by ACMS between the construction teams (Davidson and Moshini, 1990, Bowden, 2005)
Misinterpretations and miscommunications of project results and needs issues; and a centralized decision-making process and lack of delegated authority to field personnel often hindered progress and communications (Kestle, 2009)	ACMS use enables productivity enhancement of communication between project participants, reduction in project delays, alleviation of awareness of project issues among all parties, and easy access to and retrieval of project information (Alshawi and Ingirige 2003, and Stewart and Mohamed, 2004)

There is frequent shortage of materials, higher risk margins, serious delay in sorting out a number of project queries and Delays in decision-making and decisions are made autocratically, excessive costs, unfocused scope, and poor construction quality, unskilled workers, a lack of or no infrastructure (Justanyah and Sidawi, 2011, 2010a, b, 2012a,b)	ACMS use minimizes delays due to the arrival of updated drawings and documents, reduces visits to sites and traveling time to meetings, reduces time and money spent on disputes, automates repetitive routine processes, and eliminates paper reports (Alshawi and Ingirige 2003, and Stewart and Mohamed, 2004)
	ASMC have the ability to minimize the travel times of supervisors and may increase the frequency of progress, quality and safety inspections by providing project supervisors with systems that are easily applicable (Golparvar-Fard et al., 2011, Jaselskis et al. 2011).
	ACMS would help SEC's supervisors to monitor the Procurement, supply and consumption of construction materials. It would would help SEC eliminating some causes of potential delays. It helps in monitoring closely and frequently the contractors/sub-contractors and site personnel activities and construction process, performance and outcomes (Justanyah and Sidawi 2011 and Sidawi 2010a, b, 2012a, b)

Sidawi and Alsudairi

Sidawi and Alsudairi Visualization in Engineering 2014 2:3, doi: 10.1186/2213-7459-2-3

Barriers to the use of ACMS

Despite fast developments in IT and the creation of many IT applications for the construction industry, some issues still hinder the applicability of these systems to construction project management. There is a problem with regard to the diffusion of IT in the construction industry and the absorption of IT into work practices. This includes the level of strategic IT investment by construction industry firms (Alshawi et al. 2009). Other barriers include IT technical shortages, deployment of the system on an ad hoc basis, isolated project management practices, and costly systems (Alshawi and Ingirige, 2003, Nuria, 2005, Leskinen, 2006, 2008). To minimize such barriers and enable ICT adoption, the following issues should be investigated (Margherita and Petti, 2010):

- Strategy: the action plan deriving from an integrated view of organization's goals and priorities, people expectations, and potential benefits;
- People: the single individuals' attitude and the overall organizational context which impact on the level of willingness and readiness to change;
- Process: the real unit of analysis and trigger of change in terms of alternative redesign scenarios and associated impact; and
- Enablers: the potential facilitators of implementation at technological and organizational level

Nitithamyong and Skibniewski (2007) investigated the impact of project success/ failure factors on performance of WPMS. This is regarding the following aspects: project characteristics, project team characteristics, service provider characteristics and system characteristics. In respect to the project and project team characteristics, they found that strategic, time, cost, quality, Risk, and communication improvement are significantly linked: owner type, project type, project cost, project duration, and starting stage of PM-ASP development. In regards to project team characteristics, the above project factors are significantly linked with the project Parties deciding for the use of PM-ASP.

Anumba et al. (2006) found that the following structural and social impediments to the integration of ICT system: commitment from top management, continuing support for users, highly reliable systems, involvement of users, easy to use systems, effective project management, clear cost justification, well proven systems, full training, precisely defined objectives, facilities to meet specific needs, use of pilots or prototypes and organizational change (see also Grimshaw and Kemp 1989).

METHODS

The literature review have highlighted the potentials of the ACMS in a number of geographical locations around the world so it would be useful to test the useful of these systems for large companies within the context of KSA. This study examines the management of remote construction sites by large companies and how the use of ACMS would help these companies improving its' management of remote projects. The objectives of the research are:

- To find out possible management and communications problems of remote construction sites;

- To explore how these problems would be sorted out using ACMS and how ACMS would help in improving management practices; and

- To identify possible barriers to the full application of ACMS

To achieve the research objectives, it is argued that a combination of quantitative (i.e. a questionnaire) and qualitative (i.e. interviews) research tools are used. This combination was chosen because it would enable the researchers in creating and studying the visual representation of data. Also, the findings that relate to each method will be used to complement one another and, at the end of the study, to enhance theoretical or substantive completeness (Ausubel 1968). The research tools i.e. the questionnaire and interviews, were designed in accordance to the literature review, the findings of the pilot study that carried out in 2009 and the previous research on remote construction sites that has been conducted in the KSA in 2010.

The targeted population consists of contractors and companies' supervisors/engineers of Aramco (Saudi Arabian Oil Company), Royal commission of Jubail (RCJ) (petrochemical company), SABIC (Saudi Basic Industries Corporation and it a petrochemicals manufacturer), compendium of construction and consultancy companies which are working on remote sites of University of Dammam. These companies are located or have branches in the Eastern province, KSA and they have a number of remote construction sites of different sizes. This targeting method would provide feedback from the two major project's parties i.e. the owner and the contractor, which would increase the applicability of the proposed solutions regarding management and tools to both of them.

In mid-2012, phone calls were made to the managers and contractors of above mentioned companies asking them whether they would be happy to participate in the field survey. Seventy questionnaire forms were sent out to individuals who expressed their willingness to take part in the field survey. The progress of filling up the questionnaire forms was checked up by phone. These forms were filled in and returned back. However, after examining the returned forms, it was found that 23 of these are invalid as many parts have not been completed. Eventually, the total number of valid questionnaires is 47. The face to face interviews were conducted in January 2013 and these were with one consultant that represents the University of Dammam in respect of supervision of a number of construction projects for the university's, two engineers who represent a contracting company which is working on one of the University's projects, and seven engineers of various specialists i.e. architectural, structural engineering, and HVAC, who are working in the private projects department of the Royal Commission of Jubail. The total number of interviewees was ten. The researchers were not able to conduct any interview with Aramco engineers and contractors as none was willing to participate in the interviews stage.

The analysis of the results was undertaken using SPSS16 for quantitative analysis and a number of data visualization techniques were used. Mean, standard deviation, and skewness values were calculated, and the researchers inspected links between potential factors (i.e. independent variables) and dependent variables. Cross-referencing (i.e. similarities and non-similarities) technique was used to analyze the qualitative data. This technique enabled the researchers to classify the data into categories and make comparisons, and allowed rational interpretation and judgment. The following sections discuss remote construction site challenges and the potentials and barriers to the application of ACMS.

RESULTS AND DISCUSSION

Remote Construction Projects Problems

On a daily basis, the most requested information by the construction team on the remote construction site from the regional supervision office include the followings: clarification of design information, reporting QA/QC problems, contract drawings, initiation site inspections, reporting the site inspection outcomes, as-built records, and delay records. There is however, substantial delay in dealing with the following site's queries: revisions to submittals, meeting minutes, updated Schedule, clarification of Design information, as-built records, contract drawings and reporting QA/QC problems (Figure 1). Respondents said that there is a strong negative impact of the delay in providing feedback for the following remote project's queries on the project's process and performance: clarification of design information, contract drawings, updated Schedule, contract specifications, reporting QA/QC problems, and budget, and as-built records. Respondents were asked about remote construction sites' problems.

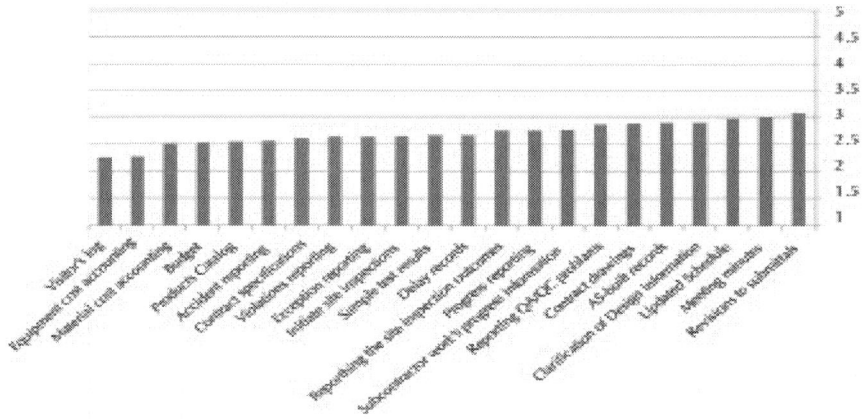

Figure 1: How frequent the respondent companies have experienced a delay in dealing with the remote construction site queries *(vertical column represents the Mean value, scale: 1 = not at all to 5 = always)*.

They said that the most frequent problems of these sites are: delay in project time table, unavailability of supervision engineers on the remote project site due to their heavy workload, the company's tendering system that obligate the choice of the lowest bidding value, the shortage of skilled labour on the site, and delay in the approval of contractor's submissions by the supervision engineers (Figure 2).

Figure 2: How frequent the following issues happen during the remote construction project's period (i.e. from start to completion). *(Horizontal column represents the Mean value, scale: 1 = does not happen at all to 5 = always happens).*

The ANOVA results showed significant links between the delay in providing information and the remote construction site problems. It showed that the more delay in providing information on the budget is associated with higher occurrence of the following problematic issues on the site: mistakes in construction works, shortage in site equipment, delay in the conduction of field survey by the contractor, unavailability of the supervision engineers during sample testing stages, improper construction methods implemented by the contractor and inadequate equipment used. The more delay in providing the products catalogue is associated with higher occurrence of ineffective planning and scheduling of the project by the contractor (Table 2). The more delay in providing the visitors log is associated with more delay in project time table. The more delay in providing revisions to submittals information is associated with higher occurrence of the use of inadequate equipment. The more delay in initiating site inspections is associated with more delay in project time table. The more delay in reporting QA problems are associated with more delay in project time table.

Table 2: ANOVA results showing correlation between the factor; *delay experienced in dealing with the budget query* with a number of dependent variables namely; *the frequency of problematic issues during the remote construction project's period*, (level of significance <0.05)

Dependent variable: *the frequency of problematic issues during the remote construction project's period (i.e. from start to completion* **Sum of Squares**		Independent variable (factor): *the delay experienced in dealing with the budget query*				
		df	Mean Square	F	Sig.	
Mistakes in construction works	Between Groups	6.105	4	1.526	2.758	0.043
	Within Groups	18.818	34	0.553		
Shortage in site equipment	Between Groups	15.838	4	3.959	3.934	0.01
	Within Groups	33.215	33	1.007		
Delay in the conduction of field survey by the contractor	Between Groups	15.123	4	3.781	5.469	0.002
	Within Groups	22.121	32	0.691		
Unavailability of the supervision engineers during sample testing stages	Between Groups	19.735	4	4.934	5.869	0.001
	Within Groups	27.739	33	0.841		
Improper construction methods implemented by the contractor	Between Groups	5.554	4	1.388	3.325	0.022
	Within Groups	13.365	32	0.418		
Inadequate equipment used	Between Groups	10.849	4	2.712	4.082	0.008
	Within Groups	22.587	34	0.664		

Sidawi and Alsudairi

Sidawi and Alsudairi Visualization in Engineering 2014 2:3, doi: 10.1186/2213-7459-2-3

The interviews revealed the following facts:

- Very few Saudi nationals are working on construction projects in KSA and there is a problem of employing non Saudis who are in KSA or recruiting HR from aboard because of the high recruitment fees on foreign contracts;

- The remoteness of remote projects has an impact of on social life of workers. The higher rank staff would find working in a remote project very difficult as they will be away from their families and the social life. The absence of electronic social communications would affect negatively the workers' productivity who are away from their families;

- Limited authority is given to the site managers and this sometimes causes substantial delay;

- Due to the cultural differences, social conflicts between workers take place and in one case, these are not reported to the project's owner or to the authorities. Also, one should be careful and choose his words when talking with staff from different backgrounds as some would understand certain words as an abuse or a threat;

- It is noticed that there are differences in productivity between workers from different backgrounds so this should be considered as it would delay the work in some parts of the project;

- all interviewed staff have QA measures but they do not apply sustainability measures;

- in one company, the contractor has weak commitment to providing frequent progress report to the supervision office as this has not been specified in the contract;

- Risk management and materials' procurement is programmed. However, each remote project has its' merits, estimation of insurance on labour, equipment, third party, risk level etc. and unexpected issues such as absence of road net or bad roads status or so. Differences in estimating the risk level exist between the contractor, the owner and other concerned parties. This affects how the project would be run by each party and may create conflict of interests between these parties;

- the government bidding and tendering system for design or/and construction of a project takes into account the lowest bidding price regardless of the quality which affect the quality of projects;

- materials were originally specified according to the specification but sometimes the materials that brought to the site did not comply with the specifications;

- one of the main causes of construction problems is the conflict between design documents that are produced by different specialists;

- the contractor should thoroughly survey the site. He should check the utilities i.e. sewage, electricity, water on the site or if there anything that would affect the project's kick-off such as the availability of road networks. SEC – for instance- did not update information of some sites so when these sites are possessed by RCJ, they may find electric cables, foundations, SEC temporary station, and septic tanks under the ground;

- Changes would happen to the original design scheme. In one occasion, the granite panels for the facades were replaced with precast concrete panels to speed up the construction, and in another incident, solar panels were included in the original design scheme, but later on these were abandoned, and the project was connected to the national electric grid. The reason was that the owner of the project was scared of trying something for the first time and considered it of high risk. In another incident, the desalination station was designed with a certain capacity. Later on, it was discovered that the capacity is not enough and should be extended to cover the present and future needs; and

- Some contractors do not have the background to manage certain types of projects and remote projects. They are incapable to manage their financial affairs, materials' procurement and how to provide qualified human resources.

The Present and Potential Use of the ACMS

The systems and tools that are mostly used by respondents for management of the projects and communications between the remote construction site and the supervision office are: e-mail through internet/ intranet/ wireless/satellite, mobile phone, laptop, site visit, weekly or monthly reports and weekly meetings. Advanced computer tools such as Web-based Project Management system (WPMS), personal digital assistants, virtual private network, and construction Robots are of little use.

Respondents said that the use of the following systems would help in efficiently manage the remote construction projects: e-mail through internet/intranet/wireless/Satellite, intelligent mobile phones, tablet computer, and Web-based Project Management System. On the other hand, they considered the following systems would help less in managing remote sites: the Walkie-talkie, web monitoring cameras and construction robots. Respondents said the ACMS would highly accelerate the feedback of the following site queries: reporting QA/QC problems, contract drawings, reporting the site inspection outcomes, clarification of design information, accident reporting and violations reporting. On the other hand, it was considered of lesser help in providing fast feedback regarding the following sites queries: material cost accounting, equipment cost accounting, and visitor's log (Table 3).

Table 3: How fast the ACMS would accelerate the feedback of the remote construction site' queries, (*Mean value scale: 1 = do not accelerate to 5 = highly accelerate*)

Type of the remote construction site's query	Mean	Std. deviation	Skewness
Reporting QA/QC problems	4.13	0.992	-0.825
Contract drawings	4.11	0.841	-0.468
Reporting the site inspection outcomes	4.06	0.942	-0.62
Clarification of Design information	4.02	0.944	-1.016
Accident reporting	3.96	1.083	-0.879
Violations reporting	3.94	1.131	-1.001
Progress reporting	3.91	1.08	-1.014
Revisions to submittals	3.89	0.983	-0.209
Initiate site inspections	3.89	1.005	-0.45
Contract specifications	3.83	1.018	-0.693
Updated Schedule	3.83	1.253	-0.72
Sample test results	3.77	1.165	-0.812
Exception reporting	3.74	1.206	-0.879
Meeting minutes	3.68	1.163	-0.72
As-built records	3.64	1.293	-0.73
Delay records	3.63	1.388	-0.752
Subcontractor work's progress information	3.48	1.11	-0.452

Products Catalog	3.36	1.342	-0.368
Budget	3.33	1.266	-0.242
Material cost accounting	3.26	1.307	-0.447
Equipment cost accounting	3.2	1.222	-0.316
Visitor's log	3.13	1.377	-0.135

Sidawi and Alsudairi

Sidawi and Alsudairi Visualization in Engineering 2014 2:3, doi: 10.1186/2213-7459-2-3

Respondents said that ACMS would be very helpful in achieving the following project lean and sustainable values: pursuing perfection by continuous improvement, specifying precisely value from the perspective of the ultimate customer/owner, identifying clearly the process that delivers what the customer/owner values (the value stream) and eliminating construction works' waste.

On the other hand, it would provide less support to managers who aim to achieve the following values: protecting the KSA people health and reducing the use of natural resources. The interviewees said that ACMS would help:

- Facilitating the procurement of materials whereas requests for materials are sent electronically to the head office for approval. However, as some materials are imported from aboard, there is a need to see the material or a sample before purchasing it.
- sending requests for specialist HR to recruitment agents who recruit HR from overseas. It also helps in mobilization of HR from a project to another as needed.
- overcoming cultural differences and understanding the other and, for instance, including religious festivals holidays in the project schedule
- providing fast social communications between the worker on site and his family at home or abroad is essential and this would bring relief to the worker. Fast communications channels are required.
- delegating certain level of authority to the sites' managers. This would help in mobilization of HR and equipment between sites; and

- providing updated benchmarks/standards/measures on the net so staff would refer to it, while examining certain case. Thus they would take the decision without getting an approval from the higher authorities

However, the management system should be transparent, so any investment opportunity by the headquarters should be known by all concerned departments.

BARRIERS TO THE USE OF ACMS

This study suggests that the above mentioned companies experience a number of unique construction management performance problems. The study found significant links between these performance issues and how far the ACMS would help in providing fast feedback to the site's queries and in achieving project's values. Therefore, these issues would affect the adoption and utilization of the ACMS; and would be considered as barriers to the use of ACMS (see also Nitithamyong and Skibniewski 2007).

The field survey show frequent construction problems and delays in responding to some site queries (see also Deng et al. 2001). Also it showed some unexpected problems i.e. infrastructure, social, psychological as mentioned above (see McAnulty and Baroudi 2010). This would raise the risk level and by turn the cost of remote projects in comparison with normal projects. The interviews show problems related to materials' procurement, transportation and installation, differences in estimating the risk level between project parties (Kestle and London 2002, 2003).

Differences in estimating cost, time, scope, and quality management between parties and the traditional management practices can be seen as obstacles to ACMS full utilization. There is lack of skilled human Resources, and there are cultural differences and a language barrier between workers and communication problems between the senior staff and labor. The cultural differences occasionally create tensions between teams (ibid).

ACMS systems are of little use by the respondents and site personnel (i.e. the contractor, sub-contractor or the owner's employees). The ACMS was considered of little help regarding the financial matters such as Budget, material cost accounting, and equipment cost accounting. Also, the ACMS would provide lesser support to managers who aim to achieve the following values: protecting Saudi Arabia's peoples' health and reducing the use of natural resources. The interviews indicated that in some companies, fragmented and standalone software was used for managing projects. The IT system of the project owner at the company headquarters is sometimes not connected with its departments and mostly not linked with the contractor or other project parties' systems. Also, problems regarding the weak IT infrastructure (e.g. slow servers) are experienced. Staff finds some difficulty in dealing with updated developments of software. In one of the companies, IT staff have good IT skills but without good construction background. None of the interviewees use Building Information Modelling (BIM) programs or receives BIM documents from consultants or design firms. All documents received are AutoCAD ones or based on AutoCAD drawings.

CONCLUSIONS: HOW TO MAXIMIZE THE POTENTIALS

To maximize the potentials of ACMS, it should be designed to provide innovative and intelligent solutions to the unique remote construction problems. The recommendations for the application of ACMS below can be generalized to other remote construction sites aboard taking into account the location and project's parameters. The design of future ACMS should consider the following issues:

The Management Side: The present researchers that design of ACMS would have the following management levels:

Strategic Planning Level: the company's strategy should be programmed into the ACMS system and this includes the strategic management of all its' remote projects in terms of how to coordinate activities, mobilization/demobilization of HR, equipment, materials, transportation, and other logistics issues.

Project Planning and Process: automated design and pre-planning of site activities is essential as this should consider the risky variables' of the environment, the remote site and project as mentioned above. There are differences in risk estimation between project parties, and this should be discussed during the design stage and the level of risk should be agreed by all parties thus embedded in the system and all contract documents.

The ACMS should provide flexible decision-making mechanisms and consider short and long-term partnering, and sharing information and management and communications tools with contractors and other project parties. This would enhance knowledge integration and help to foster innovative ideas that dramatically improve projects (Barlow 2000). During the construction stage, ACMS should provide precise daily control and follow-up procedures regarding issues such as remote examination of work quality, monitoring productivity of site workers, and calculation of material consumption rates. Some problems seem to be generated during the early stages of the project. These should be programmed as well. ACMS should enable the creation of new hypothesized cases, and the input of existing cases of emergency scenarios and the generation of possible intelligent solutions. ACMS would have an educational/training part that helps various project parties to learn about unique previous problems of the construction of remote projects and possible solutions. This would help staff overcoming present remote site's problems. The system would also have an assessment section wherein it checks whether the contractor has the ability and knowledge to run the designated remote project. ACMS should provide transparency and it should consider how to reduce the negative impacts of the project on the environment and environmental parameters on project performance and processes.

It is important to provide all the necessary information i.e. benchmarks/standards/measures on the system so it would be used by staff without the need to get the approval from the higher authority. However, financial, sustainable, and environmental aspects of the project should be considered in the design of the ACMS system.

Human Resources Level: the future ACMS would help all parties to respond efficiently to unexpected issues, to quickly sort queries, to lead multinational teams, to sort out potential social conflicts and find out possible solutions, to coordinate with other remote sites' managers and overcome communications problems. Religious festivals and holidays should be considered and frequent social events should be programmed. The system should give the managers the required level of authority and this helps in mobilization and demobilization of HR and equipment between distant projects.

Staff should be informed about the benefits and advantages of new ACMS systems. Managers should be trained on site and how to manage remote sites virtually. The design of the new ACMS in-house or adoption of new ACMS should be discussed with the contractors and supervisors/engineers to find out their views, perceptions and expectations.

The Social and Cultural Side: ACMS should help bridging cultural differences and overcoming language barriers between workers and communication problems between the senior staff and labor.

ACMS should help facilitating the procurement of materials whereas requests for materials are sent electronically to the headquarters for approval. ACMS should help in sending requests for specialist HR to recruitment agents who recruit HR from overseas. Through this system, fast social communications should be provided between the worker on site and his family at home or abroad.

AUTHORS' CONTRIBUTIONS

BS and AA together carried out the filed study. Subsequently, the gathered data was analyzed. All authors contributed to the writing of the manuscript. Both authors read and approved the final manuscript.

REFERENCES

1.   Ahuja, V, Yang, J, Shankar, R (2010). IT-enhanced communication protocols for building project management. *Engineering, Construction and Architectural Management Journal, 17*(2), 159–179.

2.   Ahuja, V, Yang, J, Shankar, R (2009). Benefits of collaborative ICT adoption for building project management. *Construction Innovation: Information, Process, Management, 9*(3), 323–340.

3.   Alshawi, M, & Ingirige, B (2003). Web-enabled project management: an emerging paradigm in construction. *Automation in Construction, 12*, 349–364.

4.   Alshawi, M, Goulding, J, Khosrowshahi, F, Lou, E, Underwood, J (2009). How strategic is IT investment in the construction

industry? A UK Perspective, Modern Built Environment – Knowledge Transfer Networks, Intelligent Buildings Index. 1–3.

5. Anumba, CEH, Dainty, A, Ison, S, Sergeant, A (2006). Understanding structural and cultural impediments to ICT system integration: A GIS-based case study. *Engineering Construction and Architectural Management*, *13*(6), 616–633.

6. Arayici, Y, & Aouad, G (2010). Building information modelling (BIM) for construction lifecycle management, in: Construction and Building: Design, Materials, and Techniques. (pp. 99–118). NY, USA: Nova Science Publishers.

7. Arayici, Y, Egbu, C, Coates, P (2012). Building information modelling (BIM) implementation and remote construction projects: Issues, Challenges and Critiques, ITcon Vol. 17, Special Issue Management of remote construction sites and the role of IT Systems. 75–92. http://www.itcon.org/2012/5 *webcite*. Accessed 10 July 2013

8. Ausubel, DP (1968). Educational Psychology: A Cognitive View. New York, NY: Holt, Rinehart & Winston.

9. Barlow, J (2000). Innovation and learning in complex construction projects. *Research Policy*, *29*, 973–989.

10. Becerik, B (2005). Critical Enhancements for Improving Adoption of OPM Technologies. Harvard Graduate School of Design, Barrie Award Winning Reports, PMI Educational Foundation Funded. http://i-lab.usc.edu/documents/Enhancements%20for%20 IT%20Adoption_PMI_Sept%202005.pdf*webcite*. Accessed 10 December 2012

11. Bowden, S (2005). Application of mobile IT in construction. University of Loughborough, Department of Civil & Building Engineering.

12. Charoenngam, C, Ogunlana, SO, Ning-Fu, K, Dey, PK (2004). Re-engineering construction communication in distance management framework. *Business process management Journal*, *10*(6), 645–672.

13. Davidson, CH, & Moshini, R (1990). Effects of organisational variables upon task organisations' performance in the building industry. *Building Economics and Construction Management*, *4*, 17–22.

14. Deng, ZM, Li, H, Tam, CM, Shen, QP, Love, PED (2001). An application of the Internet-based project management system. *Automation in Construction, 10*(Elsevier), 239–246.

15. Golparvar-Fard, M, Peña-Mora, F, Savarese, S (2011). Integrated sequential as-built and as-planned representation - with D4AR – 4 dimensional augmented reality - tools in support of decision-enabling tasks in the AEC/FM industry. ASCE Journal of Construction Engineering and Management. http://dx.doi.org/10.1061/(ASCE)CO.1943-7862.0000371. Accessed 10 April 2012

16. Grimshaw, DJ, & Kemp, B (1989). Office automation in local government in the UK. *Local Government Studies, 15*(2), 7–15.

17. Jaselskis, E, Ruwanpura, J, Becker, T, Silva, L, Jewell, P, Floyd, E (2011). Innovation in Construction Engineering Education Using Two Applications of Internet–Based Information Technology to Provide Real–Time Project Observations. ASCE Journal of Construction Engineering and Management. http://dx.doi.org/10.1061/(ASCE)CO.1943-7862.0000297 Accessed 12 July 2012

18. Justanyah, N, & Sidawi, B (2011). The dilemma of communications and management of remote construction projects in the kingdom of Saudi Arabia, Sixth International Conference on Construction in the 21st Century (CITC-VI). (pp. 395–406). Kuala Lumpur, Malaysia: Construction Challenges in the New Decade.

19. Kestle, L, & London, K In Formoso C, Ballard G (Eds.) (2002). Towards the development of a conceptual design management model for remote sites. *10th Annual Conference on 'Lean Construction (IGLC-10) (Vol. 1, pp. 309–322)* Gramado: Federal University of Rio Grande Do Sul.

20. Kestle, L, & London, K (2003). Remote site design management –the application of case study methodology. Paper presented at the Postgraduate construction research conference. Melbourne: RMIT.

21. Kestle, L (2009). Remote Site Design Management. NZ: University of Canterbury.

22. Leskinen, S (2006). Mobile Solutions and the Construction Industry Is it a working combination?.VTT, publications. http://

www.vtt.fi/inf/pdf/publications/2006/P617.pdf.   Accessed   18 February 2011

23. Leskinen, S (2008). Mobile technology in the Finnish construction industry – present problems and future challenges. 21st Bled eConference eCollaboration: Overcoming Boundaries through Multi-Channel Interaction June 15–18, 2008. Slovenia: Bled.

24. Margherita, A, & Petti, C (2010). ICT-enabled and process-based change: an integrative roadmap.*Business Process Management Journal*, *16*(3), 473–491.

25. McAnulty, S, & Baroudi, B (2010). Construction Challenges in Remote Australian Locations. Leeds, United Kingdom: Association of Researchers in Construction Management (ARCOM) Conference.

26. Nitithamyong, P, & Skibniewski, MJ (2004). Web-based construction project management systems: how to make them successful? *Automation in Construction, 13*, 491–506

27. Nitithamyong, P, & Skibniewski, JM (2006). Success/Failure factors and performance measures of web-based construction project management systems: professionals' viewpoint. *Journal of Construction Engineering and Management ASCE, 132*(1), 80–87.

28. Nitithamyong, P, & Skibniewski, MJ (2007). Key success/failure factors and their impacts on system performance of web-based project management systems in construction, ITcon Vol. 12. 39–59. http://www.itcon.org/2007/3 *webcite*. Accessed 14 January 2011

29. Nuria, FM (2005). Life cycle document management system for construction. PhD thesis. Universitat Politechica De Catalunya. http://www.tdx.cat/bitstream/handle/10803/6160/01Nfm01de12.pdf?sequence=1   Accessed 12 October 2011

30. Pena-mora, F, Vadhavkar, S, Aziz, Z (2009). Technology strategies for globally dispersed construction teams. Journal of Information Technology in Construction. http://www.itcon.org/2009/08. Accessed 5 August 2011

31. Sidawi, B (2010a). The sustainable management of remote construction projects. Fes, Morocco: Arab Society of Computer Aided Architectural Design AsCAAD.

32.   Sidawi, B (2010b). The use of advanced electronic management systems to manage remote projects in the Kingdom of Saudi Arabia. (pp. 633–642). Leeds, UK: Association of researchers in Construction management conference (ARCOM).

33.   Sidawi, B (2012a). Remote construction projects' problems and solutions: the case of SEC. ASC 48th International Conference held in conjunction with the CIB Workgroup 89. UK: Birmingham City University.

34.   Sidawi, B (2012b). Potential use of communications and project management systems in remote construction projects: the case of Saudi Electric Company. *Journal of Engineering, Project, and Production Management*, *2*(1), ISSN 2221–6529 (Print), ISSN 2223–8379 (Online), 14–22, 2012.

35.   Stewart, RA, & Mohamed, S (2004). Evaluating web-based project information management in construction: capturing the long-term value creation process. *Journal of Automation in Construction, Elsevier Science*, *13*(4), 469–479.

36.   Thomas, S, Palaneeswaran, E, Kumaraswamy, MM (2003). Web-based centralized multiclient cooperative contractor registration system. *Journal of Computing in Civil Engineering*, *17*(1), 28–37.

37.   Thorpe, D (2000). E-projects in action – the online remote construction management research project. Australia: Construction Industry Institute.

38.   Vadhavkar, S, & Pena-Mora, F (2002). Empirical Studies of the Team Interaction Space: Designing and Managing the Environments for Globally Dispersed Teams. Gaithersburg, MD: USA: NIST.

39.   Wamelink, JWF, Stoffelem, M, der Aalst, V (2002). W.M.P. International Council for Research and Innovation in Building and Construction. CIB w78 conference, Aarhus School of Architecture, 12–14 June 2002. Denmark: CIB.

40.   Yang, J, Ahuja, V, Shankar, R (2007). Managing Building Projects through Enhanced Communication – An ICT Based Strategy for Small and Medium Enterprises, CIB World Building Congress 2007. (pp. 2334–2356). South Africa: CIB.

41.   Zou, P, & Roslan, B (2005). Different perspectives towards using web-based project management systems in construction: large enterprises versus small- and medium-sized enterprises. *Architectural Engineering and Design Management*, *1*(2), 127–143

Solid Waste as Renewable Source of Energy: Current and Future Possibility in Algeria

Boukelia Taqiy Eddine and Mecibah Med Salah

Mechanical Department, Faculty of Engineering, University of Mentouri, Constantine, 25000, Algeria

ABSTRACT

Algeria has created a green momentum by launching an ambitious program to develop renewable energies and promote energy efficiency. Solid waste is one of most important sources of biomass potential in Algeria, which can be used as renewable energy sources.

With economic development and the evolution of population, the quantity of solid waste is increasing rapidly in Algeria; according to the National Cadastre for Solid Waste Generation, the overall generation of municipal solid waste was more than 10.3 million tons per year, and the amount of industrial solid waste, including non-hazardous and inert industrial waste was 2,547,000 tons per year, with a stock quantity of 4,483,500 tons. The hazardous waste generated amounts to 325,100 tons per year; the quantities of waste in stock and awaiting a disposal solution amount to 2,008,500 tons. Healthcare waste reaches to 125,000 tons per year.

The management of solid waste and its valorization is based on the understanding of solid waste composition by its categories and physicochemical characteristics. Elimination is the solution applied to 97% of waste produced in Algeria. Wastes are disposed in the following ways: open dumps (57%), burned in the open air in public dumps or municipal uncontrolled ones (30%), and controlled dumps and landfill (10%). On the other side, the quantities destined for recovery are too low: only 2% for recycling and 1% for composting.

Waste to energy is very attractive option for elimination solid waste with energy recovery. In this paper, we give an overview for this technology, including its conversion options and its useful products (such as electricity, heat and transportation fuel), and waste to energy-related environmental issues and its challenges.

REVIEW

Introduction

In order to use the enormous source of renewable energies, Algeria has created a green momentum by launching an ambitious program to develop renewable energies (RES) and promote energy efficiency. This program leans on a strategy focused on developing and expanding the use of inexhaustible resources, such as solar, biomass, geothermal, wind, and hydropower, energies in order to diversify energy sources and prepares Algeria for tomorrow.

The program consists of installing up to 22,000 MW of power-generating capacity from renewable sources between 2011 and 2030, of which 12,000 MW will be intended to meet the domestic electricity demand and 10,000 MW destined for export [1]. This last option depends on the availability of a demand that is ensured on the long term by reliable partners as well as on attractive external funding. In this program, it is expected that about 40% of electricity produced for domestic consumption will be from renewable energy sources by 2030.

Solid waste is one of most important sources of biomass potential in Algeria, which is a by-product from human activities, and is characterized by the negative impacts that may affect man and the environment when disposed in an inappropriate way without treatment. Due to the continuously increasing amount of solid waste generated, particularly in capitals and major urban centers, the challenge for the governments is to reduce the waste's harmful impacts to both health and the environment. This paper is an investigation on the possibility to use solid waste as a source of bioenergy in Algeria.

Country Profile

Algeria, situated in the center of North Africa between the 35° to 38° latitude north and 8° to 12° longitude east, has an area of 2,381,741 km²[2,3]. In the west, Algeria borders with Morocco, Mauritania, and Occidental Sahara; in the southwest, with Mali; in the east, with Tunisia and Libya; and in the southeast, with Niger (Figure 1). The geographic location of Algeria signifies that it is in a position to play an important strategic role in the implementation of renewable energy technology in the north of Africa.

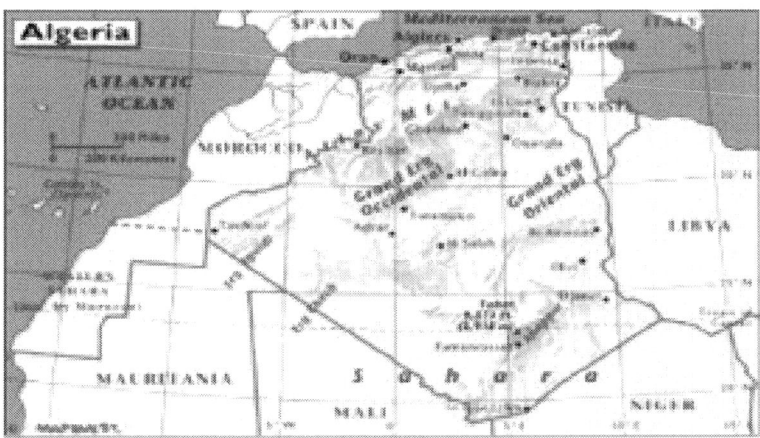

Figure 1: Map of Algeria.

The climate is transitional between maritime (north) and semi-arid to arid (middle and south). The mean annual precipitation varies from 500 mm (in the north) to 150 mm (in the south). The average annual temperature is around 12°C.

Algeria has a strongly growing population, with 36,275,358 inhabitants in 2011 according to Population Reference Bureau [4]. In the last 25 years, it has almost doubled. Though we notice a slowing down in the 1990s, the last statistics (sizable increase in marriage rate and in fertility rate) indicate that it is a short-term phenomenon, and the population growth is taking a turn towards fast growth. Algeria is characterized by a young and growing population and a fast urbanization. This situation puts certainly a lot of pressure on the energy, food supply, and even on the environment by increasing the generation of waste and residues.

Algeria plays a very important role in the world energy markets, both as a significant hydrocarbon producer and as an exporter, as well as a key participant in the renewable energy market. According to the 2011 BP Statistical Energy Survey, in 2010, Algeria had proved a natural gas reserve of 4.5 trillion m^3 and natural gas production of 80.41 billion m^3 with consumption of 28.87 billion m^3. Algeria had proved an oil reserve of 12.2 billion barrels at the end of 2010 and produced an average of 1,809 thousand barrels of crude oil per day, according to the same survey; Algeria consumed an average of 327.03 thousand barrels a day of oil in 2010 [2,3,5].

Promotion of Renewable Energies in Algeria

Algeria's location has several advantages for extensive use of most of the RES, in which it has very important potential of renewable energies including thermal solar (169,440 TWh/year), photovoltaic solar (13.9 TWh/year), and wind energy (35 TWh/year).

Solar Energy

Fortunately, Algeria has enormous potential of solar energy. More than 2,000,000 km^2 receives a yearly sunshine exposure equivalent to 2,500 KWh/m^2. The mean yearly sunshine duration varies from a low of 2,650 h on the coastal line to 3,500 h in the south.

In addition, as shown in Table 1, the potential of daily solar energy is important. It varies from a low average of 4.66 kWh/m² in the north to a mean value of 7.26 kWh/m² in the south.

Table 1: Regional daily solar energy and sunshine duration in Algeria [3]

Parameters	Region		
	Coastal line	**High plateaux**	**Sahara**
Area (km²)	95,271	238,174	2,048,296
Mean daily sunshine duration (h)	7.26	8.22	9.59
Solar daily energy density (kWh/m2)	4.66	5.21	7.26
Potential daily energy (10¹² Wh)	443.96	1,240.89	14,870.63

Eddine and Salah

Eddine and Salah International Journal of Energy and Environmental Engineering 2012 3:17, doi:10.1186/2251-6832-3-17

Photovoltaic solar energy projects in Algeria: With a potential of 13.9 TWh/year, the government plans launching several solar photovoltaic projects with a total capacity of 800 MW by 2020. Other projects with an annual capacity of 200 MW are to be achieved over the 2021 to 2030 period.

Concentrating solar thermal energy: Pilot projects for the construction of two solar power plants with a storage total capacity of about 150 MW each will be launched during the 2011 to 2013 period. These will be in addition to the hybrid power plant (solar-gas) project of Hassi R'Mel which was built for 130 MW of gas and 25 MW of thermal solar energy with the parabola system of the giant mirrors on a surface of approximately 180,000 m².

Four solar thermal power plants with a total capacity of about 1,200 MW are to be constructed over the period of 2016 to 2020. The program of 2021to 2030 provides for the installation of an annual capacity of 500 MW until 2023, then 600 MW per year until 2030.

Wind Energy

Wind is another renewable source that is very promising with a potential of 35 TWh/year. The wind map of Figure 2, established by the Centre of Renewable Energies Development (CDER) and the Ministry of Energy and Mines (MEM), shows that 50% of the country surface presents a considerable average speed of the wind. The map also shows that the south-western region experiences high wind speeds for a significant fraction of the year. The Algerian RES program plans at first, in the period of 2011 to 2013, the installation of the first wind farm with a power of 10 MW in Adrar. Between 2014 and 2015, two wind farms with a capacity of 20 MW each are to be developed. Studies will be led to detect suitable sites to realize the other projects during the period of 2016 to 2030 for a power of about 1,700 MW.

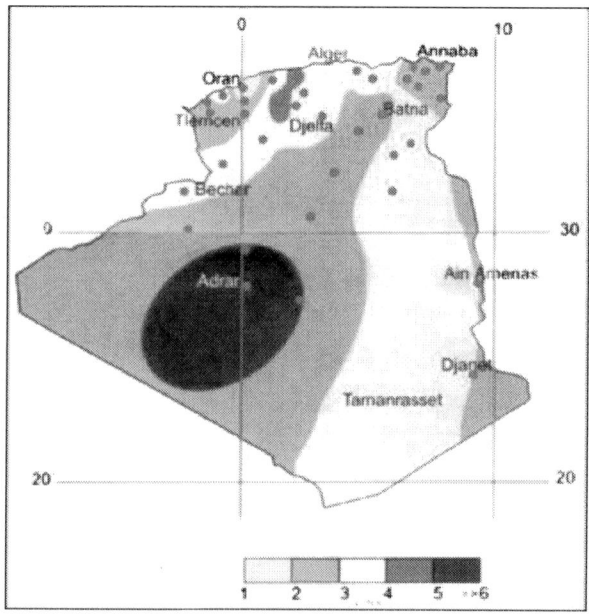

Figure 2: Wind chart of Algeria.

Geothermal Energy

The Algerian potential of geothermal energy is estimated at 460 GWh/ year [7]. More than 200 geothermal sources were counted by the CDER [7] and are recorded, of which one-third of the temperatures are superior to 45°C and where the highest temperatures registered are 98°C and 118°C in Hamam El Maskhoutin and Biskra, respectively, situated in the western part of the country. So far, the applications are limited to agricultural (heating of greenhouses, aquaculture), space heating, sanitary, and balneotherapy.

Hydroelectricity

The overall flows falling over the Algerian territory are important and estimated to be 65 billion m^3but of little benefit to the country due to the following reasons: restrained rainfall days, concentration on limited areas, high evaporation, and quick evacuation to the sea.

Schematically, the surface resources decrease from the North to the South. Currently, the evaluated useful and renewable energies are about 25 billion cubic meters, of which approximately two-thirds is for the surface resources. Hydraulic electricity represented, with 265GWh in 2003, barely 1% of the total electricity production.

Biomass Potential

The biomass potentially offers great promises with a bearing of 3.7 millions of tons of oil equivalent (TEP) coming from forests and 1.33 million of TEP per year coming from agricultural and urban wastes; however, this potential is not enhanced and consumed yet [2,6].

Regulations from the MEM which supports the use of biomass from energy crops rapidly caused an increase consumption of biomass, and in the interest in connecting the agriculture and energy sectors, this is seen as a first step in stimulating the use of biomass in Algeria much faster.

Solid Waste Generation in Algeria

In this paper, classifications of solid wastes have been proposed according to its origin into three types: municipal solid waste (MSW), industrial solid waste (ISW), and healthcare solid waste (HW).

According to the National Cadastre for Generation of Solid Waste in Algeria, the quantity of MSW generated in Algeria is estimated at 10.3 million tons/year (household and similar waste). The overall generation of ISW, including non-hazardous and inert industrial waste, is 2,547,000 tons/year with a stock quantity of 4,483,500 tons.

The hazardous waste generated amounts to 325,100 tons/year. The quantities of waste in stock and awaiting a disposal solution amount to 2,008,500 tons. Healthcare waste reaches to 125,000 tons/year according to the same source.

Municipal Solid Waste

MSW is generally defined as waste collected by municipalities or other local authorities. It includes mainly household (domestic waste), commercial, and institutional wastes (generated from shops and institutions). These wastes are generally in solid or semi-solid form. It can be classified as biodegradable waste that includes food and kitchen waste, green waste, and paper (can also be recycled); recyclable materials such as paper, glass, bottles, cans, metals, certain plastics, etc.; inert waste such as construction and demolition wastes, dirt, rocks, and debris; composite waste which includes waste clothing, tetra packs, and waste plastics such as toys; domestic hazardous waste (also called 'household hazardous waste'); and toxic waste like medication, e-waste, paints, chemicals, light bulbs, fluorescent tubes, spray cans, fertilizer and pesticide containers, batteries, and shoe polish.

According to the National Waste Agency (AND), Algeria produces 10.3 million tons of MSW each year or 28,219 tons per day, with a collection coverage of 85% in urban areas and 60% in rural areas, and a rate of 0.9 kg/inhabitant/day for urban zones and 0.6 kg/inhabitant/day for rural zones. In the capital (Algiers), the production is close to 1.2 kg/inhabitant/day [8].

The composition of MSW is closely related to the level of economic development and lifestyle of the residents. In different districts of the same city, the composition of MSW will be different. In general, the composition of MSW in Algeria with six major categories of waste was identified: organic matter, paper-cardboard, plastics, glass, metals, and others (Table 2).

Table 2: Waste composition category [17]

Waste category	Waste components
Organic matter	Waste from foodstuff such as food and vegetable refuse, fruit skin, stem of green, corncob, leaves, grass, and manure
Paper	Paper, paper bags, cardboard, corrugated board, box board, newsprint, magazines, tissue, office paper, and mixed paper (all papers that do not fit into other categories)
Plastic	Any material and products made of plastics such as wrapping film, plastic bag, polythene, plastic bottle, plastic hose, and plastic string
Glass	Any material and products made of glass such as bottles, glassware, light bulb, and ceramics
Metal	Ferrous and non-ferrous metal such as tin can, wire, fence, knife, bottle cover, aluminum can and other aluminum materials, foil, ware and bi-metal
Others	Materials from leather, rubber, textile, wood, and others such yard waste, tires, batteries, large appliances, nappies/sanitary products, medical waste, etc.

Eddine and Salah

Eddine and Salah *International Journal of Energy and Environmental Engineering* 2012 3:17, doi: 10.1186/2251-6832-3-17

Organic matter was the predominant category and represented 62% of waste collected. The other categories were represented as follows: paper-cardboard (9%), plastic (12%), glass (1%), metals (2%), and others (14%) (Figure 3). Demolition and construction wastes were not taken into account because they are disposed in uncontrolled open-air sites. The high consumption of fruits and vegetables by the city's inhabitants could explain the preponderance of organic matter in Algeria's waste.

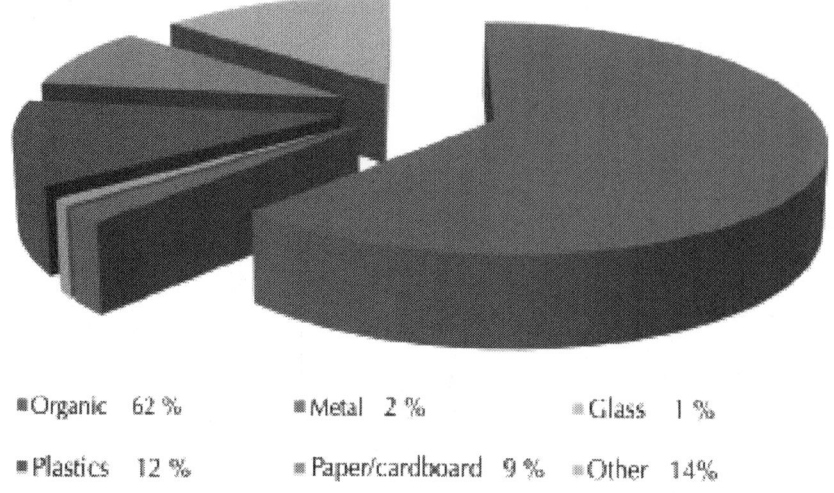

●Organic 62 %   ■Metal 2 %   ▪Glass 1 %

■Plastics 12 %   ▪Paper/cardboard 9 % ▪Other 14%

Figure 3: MSW composition in Algeria [10].

Industrial Solid Waste

According to the National Cadastre for Industrial and Special Wastes prepared in 2007, the overall generation of industrial waste, including non-hazardous and inert industrial waste, is 2,547,000 tons per year with a stock quantity of 4,483,500 tons. This type of waste is generated from the following:

Steel, metallurgical, mechanical, and electrical industries, which are the predominant sectors (50%);

Building materials, ceramics, and glass industries (50%);

Chemicals, rubber, and plastic industries (2%);

Food processing, tobacco, and match industries (29%);

Textiles, hosiery, and confection industries (10%);

Leather and shoes industries (1%); and

Wood, paper, printing industries (3%).

The hazardous waste which includes waste oil, waste solvents, ash, cinder, and other wastes with hazardous nature (such as flammability, explosiveness, and causticity) generated amounts to 325,100 tons/year. The quantities of waste in stock and awaiting a disposal solution amount to 2,008,500 tons, which are generated by four principal sectors: hydrocarbons (34%), chemistry, rubber and plastic (23%), metallurgy (16%), and mines (13%). Compared to textile (4%) as well as paper and cellulose cement and drifts, food and mechanics produce less than 2%.

Table 3 shows that the eastern regions hold the palm for the production of ISW in Algeria, with the wilayas of Annaba and Skikda which are characterized by a high proportion of waste generated and in stock (the petrochemical, transportation, and hydrocarbons industries of these regions). The western region is in the second position, because the industrial area of Arzew is the largest generator of waste with 65,760 T/year only for its refinery, followed by the industrial area of Ghazaouet with 18,500 T/year. The central region is characterized by the high production of lead waste (manufacture of battery and refinery) [11].

Table 3: Production of hazardous industrial waste in Algeria [12]

The area/production site	Production of HIW		Stock of HIW	
	The quantity (tons/year)	%	**The quantity (tons/year)**	%
The Eastern region	145,000	45	1,100,800	54.6
The Western region	98,500	30	521,800	25.6
The Central region	77,007	24	378,000	19.5
Southeast and Southwest regions	4,500	1.4	7,900	0.3
Total	325,007	100	2,008,500	100

Eddine and Salah

Eddine and Salah International Journal of Energy and Environmental Engineering 2012 3:17, doi: 10.1186/2251-6832-3-17

Healthcare Waste

These wastes include materials like plastic syringes, animal tissues, bandages, cloths, etc. This type of waste results from the treatment, diagnosis, or immunization of humans and/or animals at hospitals, veterinary and health-related research facilities, and medical laboratories. HW contains infectious waste, toxic chemicals, and heavy metals, and may contain substances that are genotoxic or radioactive. HW reach 125,000 tons/year, of which 53.6% is general waste, 17.6% is infectious waste, 23.2% is toxic waste, and 5.6% is special waste, with waste generation rate 0.7 to 1.22 kg/bed/day, in which 75% to 90% is non-clinical waste and 10% to 25% is clinical waste[13,14].

Waste Management Situation in Algeria

During the past decades, environmentally sound waste management was recognized by most countries as an issue of major concern. Waste management is an important factor in ensuring both human health and environmental protection [15].

Actors of Waste Management Services

Policy and planning: The Ministry of Land Planning and the Environment (MATE) is primarily responsible for national policy environment.

Implementation and operation: AND has the mission to support the local communities in SWM and to promote activities linked to integrated waste management.

Practice of waste management:

1. Municipalities are fully responsibility for the management and control of municipal solid waste.
2. The Ministry of the Interior and Local Communities is for financial and logistical support to the municipalities.

Control and regulatory implementation: The Directorate of Environment of each wilaya (governorate) controls and regulates the implementation of the management services.

Staff training: The National Conservatory for Environmental Training does the staff training.

Policy and Planning

Municipal Solid Waste Management National Program (PROGDEM): Launched in 2001, it has already made the development of many SWM projects (municipality master schemes, landfills, etc.) possible.

Industrial and Special Waste Management National Program: This program aims at the control and disposal of special industrial waste and potentially infectious healthcare waste.

Solid Waste Management

In general, elimination is the solution applied to 97% of waste produced in Algeria. Wastes are disposed in open dumps (57%), burned in the open air in public dumps or municipal uncontrolled ones (30%), and controlled dumps and landfill (10%) (Figure 4). On the other side, the quantities destined for recovery are too low: only 2% for recycling and 1% for composting [8].

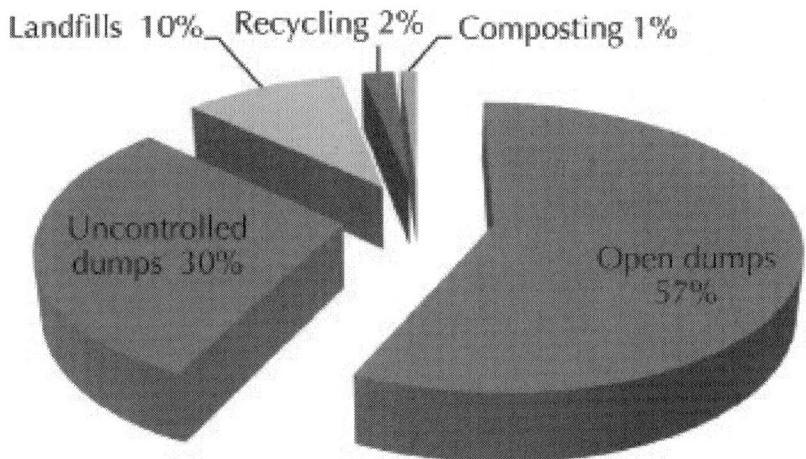

Figure 4: Methods of waste disposal in Algeria.

Open dump mode: In Algeria, the elimination of household and similar wastes through the implementation of open and uncontrolled dumps is the most common mode used, with a rate of 87%. According to an investigation by the Office of Ministry of Land Planning and the Environment, over 3,130 open dumps have been identified in the country with an area of approximately 4,552.5 ha [8]. The majority of these dumps are characterized by almost similar geographical locations. They are located along rivers, roads or agriculture lands. The other common point is that most of these dumps are almost saturated and cannot practically receive waste.

We can use the open dump of Oued Smar (Figure 5) as an actual example; it is located 13 km from the center of Algiers and was established in 1978 on an initial area of 10 ha. It now covers 32 ha and receives more than 700 trucks arriving from 56 municipalities, or 2,200 tons/day of MSW and more than 450 tons/day of rubble and fill. Table 4 provides a summary of the amount of waste placed during the period from 1978 to 2007. It went 15,000 tons during the opening year of discharge to more than 350,000 tons at the end of 2007 [8,16].

Figure 5: The open dump of Oued Smar [8].

Table 4: Quantities of waste received by the open dumps of Oued Smar (1978 to 2007) [8]

Year	1978	1980	1985	1990	1995	2000	2005	2007
Quantity (tons)	15,332	21,221	47,826	107,787	242,926	336,439	375,263	375,263

Eddine and Salah

Eddine and Salah International Journal of Energy and Environmental Engineering 2012 3:17, doi:10.1186/2251-6832-3-17

Landfill mode: Since 2001, the Algerian government has chosen to eliminate the municipal solid waste by landfill technique, which is a waste storage underground. One of the objectives of PROGDEM is to abandon the traditional mode of disposing waste by open dumps. Following the launching of PROGDEM, 65 landfills were recorded during the period from 2001 to 2005; 16 were completed, 28 under construction, and 21 during the study phase. In the end of 2007, this number has increased due to the results of pilot projects including that of Ouled Fayet in Algiers (Figure 6). It increased to 80 projects, 20 completed, 34 under construction, and 26 in study, or 15 new projects. First, the wilayas concerned are Skikda, El-Tarf, Annaba, Guelma, Souk Ahras, Batna, Tebessa, Media, Tizi-Ouzou, Setif, Biskra, Algiers, M'Sila, Ouargla, Blida, Djelfa, Jijel, Bejaia, and Chelf. In 2010, this number increased to 100 landfills, much of which was nearly completed, according to a communication of same source.

Figure 6: Landfill of Ouled Fayet [16].

The main advantages of this technology are the following:

a universal solution that provides ultimate waste disposal;

Relatively low cost and easy to implement to other waste management technology; and

Can derive landfill biogas as a byproduct for household and industrial uses.

However, this technology also has several disadvantages such as landfills requiring a large surface area and pollution problems, including ground water pollution, air pollution, and soil contamination.

The landfill of Ouled Fayet is part of the new policy of integrated waste management which included the transformation of some dumps into landfills. It serves over 34 towns of the wilayas of Algiers and Tipaza. The amount of landfill waste is 864 tons/day in 2005 against 72 tons/day when it opened in 2001. Table 5 shows the evolution of the amount of landfill waste since October 22, 2002.

Table 5: Evolution of the amount of landfill waste at the site of Ouled Fayet [8]

Period	Number of days	Number of trips	Tonnage (tons)
22/10/02 to 18/07/03	270	19,588	86,780
19/07/03 to 31/03/04	257	14,034	54,433
06/05/04 to 31/05/06	756	115,086	583,014
22/08/07 to 31/10/07	71	13,625	42,178

Eddine and Salah

Eddine and Salah International Journal of Energy and Environmental Engineering 2012 **3**:17, doi: 10.1186/2251-6832-3-17

Composting mode: Composting is a biological method for recovering organic material in solid waste. Composting represents only 1% of all waste produced in Algeria. The only experiments are those of the wilayas of Blida, Algiers, Tlemcen, and Tizi-Ouzou. The main benefit of this technology is that it converts decomposable organic materials into organic fertilizers.

We can give the example of composting station in the city of Blida; this station was put into service in 1989 and rehabilitated during the period from 1992 to 1996 and returned to service in 1996. It spreads over an area of 3.7 ha for a nominal capacity of 100 tons per shift for 8 h and for a production of 40 tons of compost.

Recovery and recycling mode: Depending on the services of the MATE, Algeria has the ability to recover an amount of waste estimated at 760,000 tons/year (Table 6), in which paper is the essential part in the possibility of recovery and recycling with a quantity of 385,000 tons/year. There are over 2 million tons of plastic packaging products in Algeria by 192 units, but only 4,000 tons are recovered (0.0002%).

Table 6: Recycling capacity

Types of waste	Quantity (tons/year)
Paper	385,000
Plastic	130,000
Metals	100,000
Glass	50,000
Various materials	95,000
Total	760,000

Eddine and Salah

Eddine and Salah International Journal of Energy and Environmental Engineering 2012 3:17, doi: 10.1186/2251-6832-3-17

Valorization of MSW

Valorization is the conversion of waste to energy, fuels, and other useful materials with particular focus on environmental indicators and sustainability goals. It is part of the larger endeavor of loop closing.

Physicochemical Properties of MSW

Knowledge of the physicochemical parameters of MSW allows the evaluation of the potentially harmful risks of pollution on the environment and human health. Also, it allows the determination of the best ways for the valorization of waste. The most important conditioning parameters for valorization are listed in the following sections [17,18].

Bulk Density: An important characteristic of biomass materials is their bulk density or volume. The importance of the bulk density is in relation to transport and storage costs.

Level of moisture: This represents the quantity of water in the MSW.

LCV: This is the total energy content released when the fuel is burnt in air, including the latent heat contained in the water vapor and, therefore, represents the maximum amount of energy potentially recoverable from a given biomass source content, or heat value, released when burnt in air.

Amount of Ash: This is the solid residue from the bioconversion in addition to other physicochemical properties such as volatile matter content, C/N ratio, and pH.

From previous studies which were carried out by Tabet [3], Guermoud [19], and Loudjani [9] show the values of physicochemical parameters as shown in Table 7.

Table 7: Physicochemical properties of MSW

Property	Value
pH	6 to 7
Medium bulk density	0.45 to 0.55 tons/m^3
Fraction of waste >40 mm	80% to 90%
Moisture content	50% to 60%
Volatile matter	40% to 60%
Ashes	15% to 40%
C/N	18 to 20
HCV	1,400 to 1,600 kcal/kg

Eddine and Salah

Eddine and Salah International Journal of Energy and Environmental Engineering 2012 **3**:17, doi: 10.1186/2251-6832-3-17

Waste-to-Energy Conversions

Energy from waste is not a new concept, but it is a field which requires a serious attention. There are various energy conversion technologies available to get energy from solid waste, but the selection is based on the physicochemical properties of the waste, the type and quantity of waste feedstock, and the desired form of energy. Conversion of solid waste to energy is undertaken using three main process technologies: thermochemical, biochemical, and mechanical extraction [20].

Biochemical conversion: Biochemical conversion processes make use of the enzymes of bacteria and other microorganisms to breakdown biomass. Biochemical conversion is one of the few which provide environment friendly direction for obtaining energy fuel from MSW. In most of the cases, microorganisms are used to perform the conversion process: anaerobic digestion and fermentation.

1. Anaerobic digestion is the conversion of organic material directly to a gas, termed biogas, which has a calorific value of around 20 to 25 MJ/Nm^3 with methane content varying between 45% and 75% and the remainder of CO_2 (biomass conversion) with small quantities of other gasses such as hydrogen.

2. Fermentation is used commercially on a large scale in various countries to produce ethanol from sugar crops. This produces diluted alcohols which then are needed to be distilled and, thus, suffers from a lower overall process performance and high plant cost.

Thermochemical conversion: Thermal conversion is the component of a number of the integrated waste management solutions proposed in the various strategies. Four main conversion technologies have emerged for treating dry and solid waste: combustion (to immediately release its thermal energy), gasification, pyrolysis, and liquefaction (to produce an intermediate liquid or gaseous energy carrier).

1. Combustion is the burning of biomass in air. It is used over a wide range of commercial and industrial combustion plant outputs to convert the chemical energy stored in the solid waste into heat or electricity using various items of process equipment, such as boilers and turbines. It is possible to burn any type of biomass, but in practice, combustion is feasible only for biomass with a moisture content <50%, unless the biomass is pre-dried.

2. Gasification process means treating a carbon-based material with oxygen or steam to produce a gaseous fuel. Gas produced can be cleaned and burned in a gas engine or transformed chemically into methanol that can be used as a synthetic compound.

3. Pyrolysis is the heating of biomass in the absence of oxygen and results to liquid (termed bio-oil or bio-crude), solid, and gaseous fractions in varying yields depending on a range of parameters such as heating rate, temperature level, particle size, and retention time.

4. Liquefaction is the low-temperature cracking of biomass molecules due to high pressure and results in a liquid-diluted fuel.

The advantage of this process, employing only low temperatures of around 200°C to 400°C, has to compete with comparably low yields and extensive equipment prerequisites to provide the pressure levels needed (50 to 200 bars).

5.   *Mechanical extraction*: It can be used to produce oil from the seeds of solid waste. Rapeseed oil can be processed further by reacting it with alcohol using a process termed esterification to obtain biodiesel, for example.

Biogas Market Options

The kind of energy produced from the biogas depends directly on the needs of the buyer, and there are three different forms: electricity generation, heat and steam generation, and transportation fuel [21].

Electricity generation: This is the most common form of energy produced in facilities constructed today.

1.   Combined heat and power (CHP) generation, also known as cogeneration, is an efficient, clean, and reliable approach to generating power and thermal energy from solid waste. By installing a CHP system designed to meet the thermal and electrical base loads of a facility, CHP can greatly increase the facility's operational efficiency and decrease energy costs. At the same time, CHP reduces the emission of greenhouse gasses, which contribute to global climate change.

2.   Fuel cell technology: Converting biogas to electricity via fuel cell technology offers significant increases in efficiency and, hence, is a highly desirable technology. Some biogas installations do exist, utilizing molten carbonate fuel cell technology; however, it is widely considered that solid oxide fuel cell technology is the most promising future technology due to its much higher power density and its applicability to a wide range of scales.

3.   Biogas engines: Biogas can be used as a motive power for the production of electricity using engines. A biogas-fueled engine generator will normally convert 18% to 25% of the biogas to electricity, depending on engine design and load factor.

4.   Microgas turbines: Small gas turbines that are specifically designed to use biogas are also available. An advantage to this technology is lower NOx emissions and lower maintenance

costs; however, energy efficiency is less than with IC engines and it costs more.

5. *Heat and steam generation*: Producing and selling heat and steam requires the existence of available industrial customers and matching the supply with their needs. It is also possible to use steam at institutional or domestic complexes.

6. *Transportation fuel*: Biogas is used as a transportation fuel in a number of countries. It can be upgraded to natural gas quality in order to be used in normal vehicles designed to use natural gas.

Waste-To-Energy Technology in Algeria

In Algeria, a little interest was given to this technology despite the important resources and the successful applications that have been made by the National Institute of Agronomy (El Harrach) and the CDER through the establishment of two experimental plants in Bechar and Ben Aknoun for the study of biogas production from cow dung. However, we must note the efforts of the Algerian government in these last years to develop this technology by upgrading the landfill of Ouled Fayet which has been put into operation in 2011. The project's main objective is the capture of the landfill's gas which contains 50% of methane (CH_4); the expected amount of emission reduction is 83,000 T equivalent CO_2/year [22].

In addition to biomass, power projects are at the feasibility study stage such as the Sonelgaz's biomass power project in the Oued Smar site, which has an installed capacity of 2 MW that can reach a peak of 6 MW from the discharge of this site, and the energy recovery plant of biogas generated in the landfill of Batna [12].

Waste-to-Energy-Related Environmental Issues

Reduction in Landfill Dumping

Landfills require large amounts of land that could be used for other purposes; incineration of solid waste can generate energy while reducing the volume of waste by up to 90%.

Reduced Dependence on Fossil Fuels

With advanced technologies, waste can be used to generate fuel that does not require mining or drilling for increasingly scarce and expensive non-renewable fossil-fuel resources.

Reduced Greenhouse-Gas Emissions and Pollution

Using waste as a feedstock for energy production reduces the pollution caused by burning fossil fuels. While traditional incineration still produces CO_2 and pollutants, advanced methods such as gasification, pyrolysis, and liquefaction, have the potential to provide a double benefit: reduced CO_2 emissions compared with incineration or coal plants, and reduced methane emissions from landfills.

WTE Provides Clean Energy

The WTE technology has significantly advanced with the implementation of the Clean Air Act, dramatically reducing all emissions.

Waste-to-Energy Challenges

Lack of Versatility

Many waste-to-energy technologies are designed to handle only one or a few types of waste (biomass, solid waste or others). However, it is often impossible to fully separate different types of waste or to determine the exact composition of a waste source. For many, waste-to-energy technologies to be successful, they will also have to become more versatile or be supplemented by material handling and sorting systems.

Waste-Gas Cleanup

The gas generated by processes like pyrolysis and thermal gasification must be cleaned of tars and particulates in order to produce clean and efficient fuel gas.

Conversion Efficiency

Some waste-to-energy pilot plants, particularly those using energy-intensive techniques like plasma, have functioned with low efficiency or actually consumed more energy than they were able to produce.

Toxic materials include trace metals such as lead, cadmium and mercury, and trace organics, such as dioxins and furans. Such toxins pose an environmental problem if they are released into the air with plant emissions or if they are dispersed in the soil, allowed to migrate into ground water supplies, and work their way into the food chain. The control of such toxins and air pollution is the key feature of environmental regulations governing MSW-fueled electric generation.

Regulatory Hurdles

The regulatory climate for waste-to-energy technologies can be extremely complex. At one end, regulations may prohibit a particular method, typically incineration, due to air-quality concerns, or classify ash byproducts of waste-to-energy technologies as hazardous materials. At the other end, while changes in the power industry have allowed small producers to compete with established power utilities in many areas, the electrical grid is still protected by yet more regulations, presenting obstacles to would-be waste-energy producers.

High Capital Costs

Waste-to-energy systems are often quite expensive to install. Despite the financial benefits they promise due to reductions in waste and production of energy, assembling the financing packages for installations is a major hurdle, particularly for new technologies that are not widely established in the market.

CONCLUSIONS

This paper gives an overview on the Algerian potential of solid waste including MSW, ISW, and HW as biomass sources. The management of solid waste (MSW) and valorization is based on the understanding of MSW composition by its categories and physicochemical characteristics.

Energy from waste is not a new concept, but it is a field which requires a serious attention. There are various energy conversion technologies (thermochemical, biochemical, and mechanical extraction) to produce useful products (electricity, heat, and transportation fuel).

In general, the government should, first and foremost, implement its own decisions and work towards encouraging independent renewable energy producers, in general, and energy generation by WTE technologies, in particular. By doing so, the overall energy generation capacity will increase, the dependence of Algeria on imported fossil fuels will be reduced, and a significant reduction in pollution and greenhouse gas emissions will occur.

Recommendations

The recommendations of this research are the following:

Solid waste can be used as an energy source in Algeria. However, the WTE facilities must operate under strict standards, which will minimize environmental impact and adhere to the precaution principle.

Implementation of landfill disposal techniques should be encouraged for the valorization of biogas.

Waste-to-energy and valorization of Algerian solid waste is a new subject that needs to be developed.

AUTHORS' CONTRIBUTIONS

BTE co-supervised the work, collected the references and the informations, and drafted the manuscript. MMS co-supervised the work and corrected the draft manuscript. All authors read and approved the final manuscript.

REFERENCES

1. Sonelgaz Group Company:

2. Stambouli, AB: Algerian renewable energy assessment: the challenge of sustainability. Energy Policy. 39, 4507–4519 (2011).

3. Stambouli, AB: Promotion of renewable energies in Algeria: strategies and perspectives. Ren. Sust. Energy Reviews. 15, 1169–1181 (2011).

4. Population Reference Bureau (PRB): World population data sheet (2011) http://www.prb.org*webcite* (2011). Accessed 2011

5. BP Statistical Review of World Energy: Statistical review, London

6. Boudries, R, Dizene, R: Potentialities of hydrogen production in Algeria. Int. J. Hydrogen Energy. 33, 4476–4487 (2008).

7. Himri, Y, Arif, SM, Boudghene, SA, Himri, S, Draoui, B: Review and use of the Algerian renewable energy for sustainable development. Ren. Sust. Energy Reviews. 13, 1584–1591 (2009).

8. Djemaci, B, Chertouk, MAZ: La gestion intégrée des déchets solides en Algérie. Contraintes et limites de samise en œuvre. International Centre of Research and Information on the Public

9. Loudjani, F: Guide des techniciens communaux pour la gestion des déchets ménagers et assimiles. The Ministry of Land Planning and the Environment (MATE)

10. Gourine, L: Country report on the solid waste management: Algeria. The regional solid waste exchange of information and expertise network in Mashreq and Maghreb countries

11. Ouzir, M: Gestion ecologique des déchets solides industriels: casd'étude la villed'arzew, University of M'sila (2008)

12. Louai, N: Evaluation énergétique des déchets solides en Algérie, une solution climatique et un nouveau vecteur énergétique, University of Batna (2009).

13. Bendjoudi, Z, Taleb, F, Abdelmalek, F, Addou, A: Healthcare waste management in Algeria and Mostaganem department. Waste Manage. 29, 1383–1387 (2009).

14. Sefouhi, L, Kalla, M, Aouragh, L: Health care waste management in the hospital of Batna City (Algeria), Paper presented at the Singapore International Conference on Environment and BioScience, Singapore (2011)

15. Redjal, O: Vers un développement urbain durable: Phénomène de prolifération des déchets urbains et stratégie de préservation de l'écosystème: exemple de Constantine, University of Mentouri Constantine (2005).

16. Kehila, Y, Mezouari, F, Matejka, G: Impact de l'enfouissement des déchets solides urbains en Algérie: expertise de deux Centresd' Enfouissement Technique (CET) à Alger et Biskra. Revue Francophone d'Ecologie Industrielle Déchets, Sciences & Techniques. 56, 29–38 (2009)

17. Alamgir, M, Ahsan, A: Characterization of MSW and nutrient contents of organic component in Bangladesh. EJEAFChe. 6(4), 1945–1956 (2007)

18. McKendry, P: Energy production from biomass (part 1): overview of biomass. Bioresour Technol. 83, 37–46 (2002).

19. Guermoud, N, Ouadjnia, F, Abdelmalek, F, Taleb, F, Addou, A: Municipal solid waste in Mostaganem City (Western Algeria). Waste Manage. 29, 896–902 (2009).

20. McKendry, P: Energy production from biomass (part 2): conversion technologies. Bioresour Technol. 83, 47–54 (2002).

21. Münster, M, Lund, H: Use of waste for heat, electricity and transport–Challenges when performing energy system analysis. Energy. 34, 636–644 (2009).

22. Tabet, MA: Types de traitement des déchets solides urbains: evaluation des coûts et impacts sur l'environnement. Rev. Energ. Ren. Special number Production et Valorisation –Biomasse. 2001, 97–102 (2001)

A Haitian 'Ecodistrict:' Conceptual Design for Integrated, Basic Infrastructure for the Commune of Léogâne

Hillary Brown[1] and Miriam N Ward[2]

[1]The Spitzer School of Architecture, the City College of New York, 141 Convent Avenue 3 M09, New York, NY 10031, USA

[2]Sustainability in the Urban Environment Program, The City College of New York, 160 Convent Avenue, New York, NY 10031, USA

ABSTRACT

Background

The island nation of Haiti poses an ongoing challenge to the international development community. Political instability and lack of basic public services, combined with weak technical and financial capacity, collectively undermine both philanthropy and international investment. Haiti's tragic vulnerability to natural disasters (earthquakes and hurricanes) and the public health crises that ensue have only compounded its level of destitution. These complex, interconnected circumstances doom single-threaded solutions.

Methods

The problems call instead for integrative responses, answers that cut across sectors and address multiple problems simultaneously. This proposal for critical services development in the Haitian commune of Léogâne seeks to reduce much of its poverty, malnourishment, pollution, health problems, flooding, deforestation, topsoil erosion and loss of biodiversity.

Results

The territory's long degraded ecosystem services, once regenerated through reforestation and riverbank stabilization, will boost agricultural production. More significantly, they will help provision targeted new energy, water and waste management systems designed together with synergistic exchanges in mind.

Conclusions

Across these sectors shared resource and energy flows will improve efficiencies, lower costs and reduce environmental impacts.

BACKGROUND

Set against decades of political volatility and economic stagnation, Haiti's marginal investments in critical infrastructural services have left its people struggling today with growing population pressures and increased needs in a context of dwindling resources. Haiti was already the poorest country in the Western hemisphere, with an unemployment rate as high as 70%, Adelman (2011) when its capital was devastated by a magnitude 7.0 earthquake on January 12th of 2010. The tragedy was compounded by non-existent building codes and shoddy construction: somewhere between 230,000 to 316,000 Haitians lost their lives beneath collapsing structures, with hundreds of thousands more injured (O'Connor 2012). Port-au-Prince, which houses major government and administration offices, was devastated. With so many of the nation's public servants buried in imploding buildings, basic services were disabled for the long months that followed. Globally, well-meaning nongovernmental organizations (NGOs) rushed to offer services and to rally a hobbled people. But Haiti, already a nation with the highest number of NGOs per capita, was overwhelmed by what became a disorganized aid effort. Miscommunications and the lack of coordination that ensued only added to the chaos (Klarreich & Polman 2012). The government worked with international organizations to clear rubble and restore necessary services, while trying to rebuild internal capacity with limited funds. In October 2010, the unthinkable happened again, when a cholera outbreak swept through regions lacking energy, sanitation and secure potable water. As of May, 2013, the epidemic had claimed more than 8,000 lives (Cholera in Haiti). Ambitious plans to "Build Back Better," (disaster recovery plans undertaken by the Clinton Foundation, along with many others) have unfortunately stagnated. At this writing, three years following the quake, the Haitian people find themselves with only spotty improvements, increased need, and declining opportunities (Sontag 2012).

Current Context

At the epicenter of the earthquake, the commune of Léogâne (population then, 130,000) sustained severe losses, with an estimated 93% of its buildings damaged (Eberhard et al. 2010) (Figure 1). While rubble has been substantially cleared, the few services that existed remain un-restored. Critical infrastructure needs to be comprehensively rebuilt. This poses an opportunity to consider progressive paradigms for utilities' integration and management.

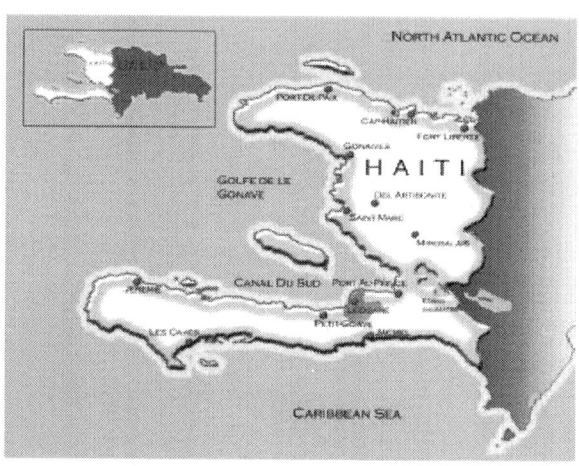

Figure 1: Haiti overview.

Official waste collection in Léogâne is non-existent. Most refuse is piled and periodically set on fire, a form of 'local management (University of Notre Dame 2011a).' Plastics, along with other household debris, are thrown askance and washed downstream, clogging drainage channels and canals. The town lacks sanitary systems for human and animal waste, and waterways are overrun with organic material, creating serious health hazards (Archibold 2012).

The existing electrical grid (servicing only a small portion of the community) has not functioned since the earthquake. While the government has promoted a plan to connect Léogâne to the national grid, nothing has transpired to date (University of Notre Dame 2011b). In July 2012, Electricité d'Haïti (EDH) and the Ministry for Public Works partnered with Korea International Cooperation Agency (KOICA) to install a small microgrid serving 20,000 Léogâne residents, powered by diesel generators (Haiti - Energy 2012). The system is still not yet fully functional. Clinics, organizations and families with resources are left to run private generators with expensive, imported fuel. Few have been able to purchase solar panels. Throughout Haiti, lack of affordable energy limits employment, buying-power, access to education, opportunities, and investment capacity (Barjon 2011a). For a nation largely comprised of remote rural poor, a distributed energy plan realized with local natural resources could greatly improve the standard of living for much of Haiti's population (OECD 2006a).

The water distribution in Léogâne is similarly problematic. During the 2008 hurricane season, the underground water supply system serving the town center was damaged and, without funding for repair, it has not functioned since (Piasecki 2013). Much of the population still struggles to access potable water. Those with means utilize private wells, while those without utilize shallow hand-dug wells, walk to a public water station or buy water in costly plastic pouches or other containers which are subsequently discarded. While the local water authority, DINEPA, has plans for restoring supply pipelines to the center, service to peripheral and rural areas will be limited to water kiosks (Rebuilding Léogâne Planning Workshop 2011a). The drainage system had been in a similar state of disrepair, but has recently been repaired through international aid. Much of the system, however, lies below sea level. Its drainage cannot be discerned, nor are the parties involved sharing operational plans (Piasecki 2013). Without transparency and local involvement progress is limited.

Transportation services are minimal and crude. Léogâne's limited roadway system is in severely deteriorated condition (University of Notre Dame 2011c). Market activities choke the town's central streets, limiting vehicle volume and speed (Shepard 2010). Roadbeds, rutted with puddles, have become vectors for mosquito nesting and water-borne illnesses. (Léogâne is a known center for mosquitos carrying lymphatic filariasis, which causes Elephantiasis.) Rural roadways are unpaved and intermittently overwhelmed by deluges, increasing the extent of erosion. To manage flooding, open drainage ditches flank roadways, posing significant hazards to unwary drivers. The lack of safe roadways also limits deployment of heavy equipment.

Large areas across Haiti have experienced severe environmental degradation. Overall, only 2% of the country remains forested (Library of Congress –Federal Research Division 2006). Both upland trees and coastal mangroves in this energy-impoverished landscape have been widely harvested for charcoal, the primary cooking fuel. Severe deforestation is also endemic in and around Léogâne, along the coastal areas and into its surrounding hills and mountains. No longer stabilized by root systems, upland soil erodes during tropical downpours. It overwhelms rivers and streams, fanning out downstream as it washes into the ocean. Routine flooding takes out roadbeds and creates impassable conditions across the alluvial plain in and around the town center.

Tropical floods have decimated the once confining banks of the Momance and Rouyonne Rivers, severely damaging not only riparian habitat, but also the deltas and estuaries. While natural causes exist for coastal habitat degradation (including the earthquake's seismic uplift that drained tidal marsh, exposed sub-tidal sea grass causing loss of coral reef and mangrove habitat), (Koehler & Mann 2011) anthropogenic impacts compound the damage through the massive influx of river sedimentation and the harvesting of mangroves for fuel (Rebuilding Léogâne Planning Workshop 2011b). Sited within the Caribbean's well-trafficked "Hurricane Alley," Léogâne's developed land area, along with its once bio-rich estuaries and coastal coral reefs, has become especially vulnerable to rising sea levels. Earthquake damage, coupled with environmental degradation, has triggered negative feedback loops, limiting local economic recovery. Practical opportunities for combined solutions to these interconnected problems have driven the integrative design of this proposed ecodistrict. (Ecodistricts are an urban planning term for sustainable development in a larger region where the ecological footprint is considered in planning and resources are used judiciously (Metzger and Olsson 2013)).

New Paradigms for Utilities in Developing Nations

Sixty percent of global energy growth in the next few decades will take place in developing nations, where about two billion people are living without access to electricity (Holm & McIntosh2008). Many emerging economies, while constrained by other substantial barriers to progress, enjoy a potential window of opportunity to bypass petroleum-dependent technologies for next generation systems. Through implementation of largely renewable energy-driven solutions upfront, these nations may accelerate the technology application in holistic ways. For example, smart-grid and distributed micro-grid networks, compared to centralized systems, reduce losses while better matching load with demand (Gregory 2011a). By investing in renewable energy sources, developing economies may draw roadmaps for a lower carbon, global energy transition. This opportune, but non-traditional pathway, however, places significant burden of risk on energy policy decision-makers. Leapfrogging fossil fuel technologies will also entail considerable support and international cooperation from both governmental and private investment communities (Holm 2005). Globally, carbon-based energy systems have enjoyed inter-governmental protection, subsidy policies, and economies of scale achieved through investment in conventional generation and transmission (Holm & McIntosh 2008).

GEMi and a Vision for Haiti

Project objectives and strategies for this ecodistrict design build upon, and are aligned with, the precepts of the Global Energy Model (GEM). The Global Energy Model Institute, a new nongovernmental initiative that bears its name (GEM Institute or GEMi), was formed to address many of these challenges (Global Energy Model Institute 2012). Its broad objective is to solve global energy poverty through the progressive application of clean, reliable power systems, based on proven technologies. GEM is the organizing framework for multi-sector investment and development with energy at its hub. Haiti is its initial pilot. A premise for its universal application is the nature of complex, cross-sectoral interactions around, and dependencies upon, energy. While GEM is technologically agnostic, it is biased towards solutions that serve the triple bottom line. The energy model itself is an integrated information management system that captures innovations and best practices and allows for successive refinement across different locales (Under development by Daniel C. Gregory of Positive Energies, LLC).

For Léogâne, it was assumed that many of GEMi's basic operating principles and implementation processes would be applied as givens. These include the reliance on stakeholder-driven development, development of Haitian professional and administrative capacity, and the utilization, wherever possible, of a local work-force. GEMi's mission also entails enabling and locally sustaining an expanding power system infrastructure through its planned education and training programs (Gregory 2010).

Haiti lacks a national grid beyond the one that currently serves Port-au-Prince and immediate environs. One of the visions for Haiti elaborated upon through GEMi involves the development of a national grid based significantly on existing hydropower that will be upgraded and enhanced (Brown 2011). For locales off the planned primarily North–south grid, a decentralized generation system is envisioned, with eventual grid connectivity. (Lemons, 2012). With Haiti's island context, renewable energy generation options are ultimately more cost-effective and viable due to the logistical and economic complications of foreign fossil fuel delivery and supply.

GEMi's reliance on diversified and networked renewable resources reduces dependence on imported fuel, limits carbon emissions and provides system diversification and resiliency. Currently, Haiti's energy is largely sourced from foreign fossil fuel hydropower; however, the existing dams are undergoing upgrades to increase the grid's base-load capacity. Refurbishing the dams risks extending negative impacts to the remaining intact ecological services contributed by the river. Sedimentation from unmanaged upland erosion must be constantly removed to maintain capacity. Moreover, methane, a highly potent greenhouse gas, is also emitted by untended decaying biological matter. Other options pose technical challenges. Both wind power and solar energy production fluctuate. Neither alone is sufficiently reliable, and each would require some measure of storage to be able to release energy on-demand. New multipronged or hybrid systems such as blended renewables, coupled with storage and backed up by diesel-fueled generation are envisioned to address the breadth of the challenge.

METHODS

Holistic Design for Synergies across Sectors: Léogâne as Model for Rural Utilities

Plans for Léogâne's redevelopment include environmental remediation and optimization of material and resource flows across the energy, water, agricultural and transportation sectors. Investment in one sector, per the plan, creates the stability needed to initialize another; this is integral to the final outcome and can't be undertaken successfully unless the supporting sectors are in place. The new paradigms for the plan rely upon the cascading of energy, water or residual resources extracted from nature across these sectors. Through colocation and coupling of complementary components, the program will capitalize on designed-in exchanges among multiple system elements. For instance, floodwater control may also be used to generate energy, irrigate crops and improve water quality. Agricultural and municipal solid waste can provide energy and heat for industrial processes while returning beneficial soil amendments. The vision entails closing loops to create a self-sufficient, holistic system, one that captures synergies, creates economic efficiencies, and eliminates waste. Promoting synergies between economic gains and social infrastructure helps to amplify benefits, ensuring that real gains are made in all sectors for the poorest and most vulnerable people (OECD 2006b). The design also promotes colocation of complementary entities such as industry, hospitals, clinics and schools, allowing these entities to take advantage of infrastructural outputs and/or contribute inputs. Overall, the plan reduces reliance upon most 'first world' carbon-intensive systems, substituting for the most part technologies that can, after training, be constructed and managed by local labor. In addition to new jobs, the scheme also entails alternative employment opportunities for agriculture, agroforestry, and aquaculture industries envisioned for the commune, along with other forms of local enterprise.

Léogâne is an ideal locus for reconstruction as its landscape mirrors that of much of Haiti, with its steep rural mountains, rivers and streams, and its populous alluvial plains. Successes in Léogâne could be used as templates for integrated development across rural Haiti where conditions are similar. Interventions in Léogâne were therefore designed along just such a typical corridor running from the mountainous uplands in the South, through the more populous alluvial plain and down to the coast on the North. The area is bounded on either side by the Momance River and the Rouyonne River, both of which have large seasonal variations.

Local efforts to guide reconstruction along this mountain-to-coast transect will help to inform placement of the system's modular elements. Working components will be sited at locations with specific relationships to both natural resources and existing settlements (Figure 2). For example, the upland placement of wind turbines captures higher wind speeds. Photovoltaic arrays, interspersed with agriculture, occupy the middle range and power the pumping of water from a nearby receiving reservoir to a holding reservoir above for the pumped-hydro-storage generation system. Within this region too, biomass salvaged from agriculture will be used to supplement energy generation. All along this corridor, riparian resources will be stabilized for flood control, infrastructure protection, and for improved water quality for farming and aquaculture.

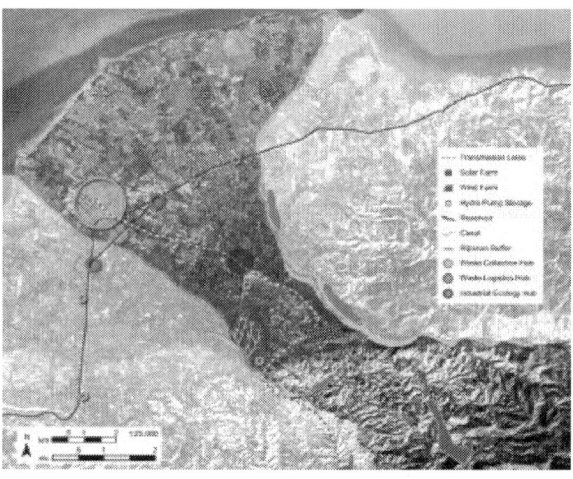

Figure 2: Integration of critical infrastructure overview.

A base load generated by a pumped-hydro storage system (PHS) will power the industrial, civic and commercial sector. Explained below, the PHS is a generation system networked for reliability and redundancy. The distribution of electrical services also relies on an integrated delivery mechanism: a set of service points or "nodes" (Figure 3). This approach standardizes and maximizes the reach of new services. In the town center, for example, nodes serve neighborhoods, public plazas, the markets and other civic functions. Nodal locations will be linked by a newly paved road network. These stations will also serve as outlets for clean drinking water and as drop-off locales for waste collection. Here, cellphones may be recharged, as well as auxiliary LED lamps. Organic and solid waste dropped off at these service points (incentivized through rebates) will be carted to a nearby waste processing (biodigestion) plant site for augmentation of the power-supply. Economies obtained from colocation and simultaneous construction of these services will make the best use of limited resources until more are at-hand to finally extend services to individual dwellings. The nodal service points create an integrated web of services linking new and existing infrastructure. A few are designed to incorporate auxiliary community services such as "electronic cafes." A number of "nodes as community-hubs" are also planned for the town square, on school campuses, clinics and at the hospital.

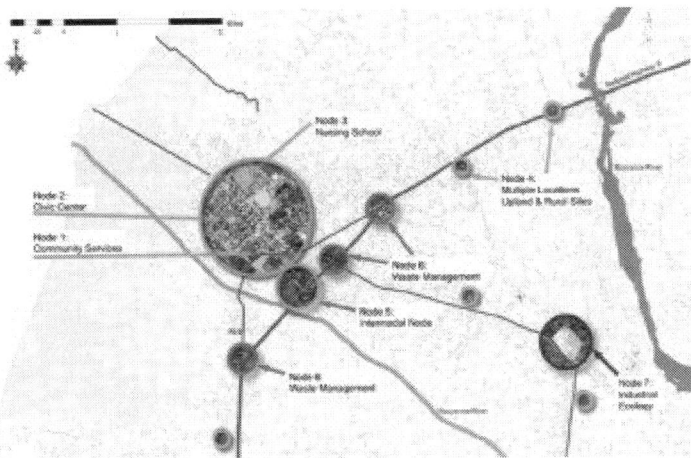

Figure 3: Downtown infrastructural nodes.

Proposal Components

Generation of Renewable Energy

Léogâne's renewable power system (with diesel back-up) will have its base load generated by PHS, powered by distributed wind and solar PV farms of 10 MW and 8.6 MW respectively. Electricity produced by these renewables will pump a quantity of water from the lower to the upper reservoir, where it is stored until released on demand (Figure 4). Passed through turbines, it generates the base-load for the microgrid. The PHS effectively acts as the storage battery and load-leveling device for the overall system (one of GEMi co-founder Daniel C. Gregory's many primary organizing concepts). Léogâne's PHS system is sized at a 10 MW, with a 180-meter head. The upper and lower reservoirs each have a water storage volume of 157,000 m³. The closed-loop flow of PHS water through surface-mounted pipes will not interfere with local riverine ecology; however, these must be placed in stabilized areas. For 8–10 hours during the day, electricity will reliably feed local micro-grids serving key industrial, commercial and residential locations, as well as some outlying areas. This type of system, which minimizes transmission losses, could serve much of Haiti where mountainous regions and streams are adjacent to urbanized areas.

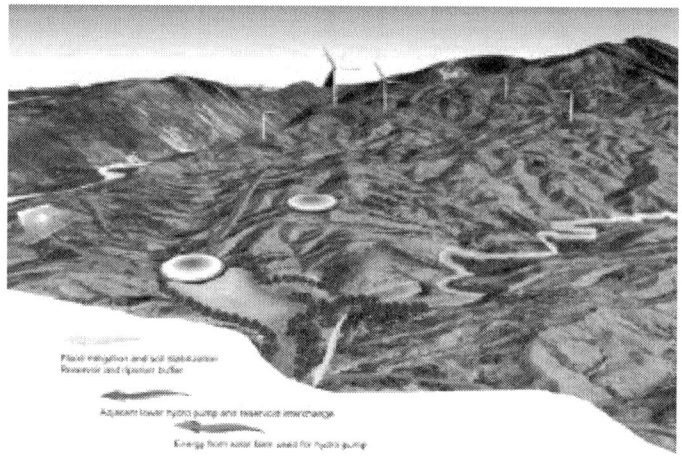

Figure 4: Energy system co-location benefits.

For system resiliency, the PHS system is reliant on multiple, complementary sources with intermittent energy supply. While Haiti's North and West Departments enjoy dependable wind velocities, Léogâne's flatlands do not reliably produce sufficient wind speeds. Placement of wind farms at higher inland elevations in the commune will better take advantage of seasonal and diurnal on- and off-shore breeze. One or more solar farms comprised of commercially available PV panels will provide electricity to the PHS during the 11–13 hours of daylight. Excess power, not needed for water pumping, will augment the microgrid. Floating PV arrays are placed spanning across each of the PHS water reservoirs, comprising part of the generation system. Floating on pontoons, these arrays shade the water, reducing evaporation. The water in turn lowers the arrays' temperatures, improving their efficiency. The arrangement also avoids unnecessary displacement of agricultural land.

Placed on opposite hilltops at elevations of 600+ meters, a total of eighty-six 100 kW wind turbines will complement the solar farm as primary feed for the pumped-storage system. These medium size turbines, selected in part for easier transport on primitive roads, are supported by lattice towers that provide greater stability in this earthquake- and hurricane-prone region.

The GEM model assumes a long-term phased approach to 24/7 power. The electrical consumption planned per household is calculated initially as quite minimal; it covers basic energy services, allowing for night illumination, and the operation of small pumps for drinking water, radio and other small appliances including cell phone charging. Community nodes and hubs are more energy intense and require power sufficient for small refrigerators, laptop computers, basic medical equipment, and some income producing productive uses. In all, it is estimated that basic household services, along with commercial, community activities (schools and clinics) could be provided for on average just 50 kilowatt-hours per person per year, (Dilip Ahuja et al. 2008) for an annual project total of 5,000 MW hours.

Telecommunications Upgrades

Even before the 2010 earthquake, Haiti's telecommunications infrastructure was generally underdeveloped (fixed line penetration at the lowest in Latin America and the Caribbean). However, the country's internet connectivity is comparatively robust as most Haitian internet service providers connect to the internet via satellite and are not reliant on undersea fiber optic cables. Also, despite the national poverty level, mobile phone coverage has grown rapidly. For that reason, the community nodes will provide capacity for cell-phone recharging and servicing as well as internet cafes.

Ecological Restoration and Enhancement

That valuable natural resources lie within its boundaries is a condition hardly unique to Léogâne. Just as elsewhere across Haiti, these ecosystem services represent one of its greatest assets. These, unfortunately, have been subject to degradation. Therefore, environmental stabilization and resource regeneration are necessary preconditions to the development of energy systems, agricultural improvements, industry and economic development. Stormwater management, flood control, soil stabilization and reforestation are critical prerequisites to the installation of the energy corridor. A program of river stabilization (repair of eroding banks and other flood control measures constructed with local materials) will alleviate hazardous deluges that occur during tropical downpours. New reservoirs located upland will store stormwater for irrigation. Riparian buffering, described below, will help reinvigorate damaged marginal areas, while the introduction of new agriculture, agroforestry, aquaculture practices, and the new rural settlements supporting them, will stabilize the topography and restore hydrological systems.

The Delta

Restoration of the imperiled coral reefs and mangroves, along with flood mitigation and bank restoration, is imperative to re-provisioning ecosystem services. Mangrove restoration must take place first so as to protect fragile corals implants that act as filtering systems, trapping debris and silt, balancing nutrient loads. Mangroves protect coastal zones from the impacts of major weather events. Reforestation and riverbank reconstruction will further reduce outflows of sediments and must be advanced with the delta restoration. Secondly, coral reefs, which take decades to build, have a vital interconnected role and must be quickly cultivated to also serve as storm surge barriers, while they rebuild local biodiversity. A promising artificial platform system called the "Reef Ball"™ is recommended for reef regeneration (The Reef Ball Foundation-Designed Artificial Reefs). It consists of a perforated concrete form with plugs of healthy new coral fragment inserts placed on the ocean floor in suitable habitat which can be economically utilized to speed the repopulation process. Mangroves can be planted in biodegradable baskets using a range of salt-tolerant and saline resistant species based on local conditions.

Riparian Buffers

The riparian corridors along the Momance and Rouyonne were former oases of ecological diversity. Stripped of native vegetation, the eroded riverbank must be rebuilt to confine river flow. The middle zone, bordering the mountainous highlands and the alluvial plain, is ill-equipped to manage flooding, resulting in soil loss from agricultural lands and damage to human settlements. Riparian restoration within a demarcated corridor between 50 to 300 feet wide will recreate vegetated and forest buffer strips and patches that armor the riverbank to absorb floodwaters and slow erosion. Through pollutant removal and temperature moderation downstream, marine biodiversity and water quality will be restored. These new riverine habitat corridors will also aid in species migration across adjacent developed landscapes. During initial restoration, land holdings must be restructured along these rivers, including no-trespassing conservation zones and a less stringent long-term management system developed with local input.

Reforestation and Agroforestry

In addition to the soil retention and ecosystem services returned through restored riparian buffers, trees and plantings are reintegrated using 'agro-forestry' models. Proposed intercropping of trees with food plants will foster stewardship while increasing land productivity through crop-shading and nutrient exchanges. Fruit production, timber from jatropha trees, vertiver grasses and coir industries can be integrated with adjacent irrigation channels to boost agricultural productivity and local enterprise. The integration of agroforestry is both economically and ecologically vital for regeneration (Collier 2009). Local inhabitants must be able to benefit while enabling ecological services to flourish.

Upland Reservoirs

New reservoirs will be constructed at higher elevations in the energy corridor for flood mitigation and water storage. Independent of the pumped-hydro storage system, these reservoirs (875 m^3 and 220 m^3 capacity for the Momance and Rouyonne Rivers respectively), will divert floodwaters. Freshwater drawn from this impounded source will be also used for local crop irrigation. Micro-turbines placed within these channels will provide power to local farmers in rural areas not serviced by the micro-grid. New 100-foot riparian buffers within this upland zone, containing an estimated 380,000 trees, shrubs and grasses combined with riprap from concrete debris, will stabilize the soil.

Other Integrated Services

Transportation

As part of the integrated services development, improvements to roadbeds and associated stormwater drainage are vital to support the installation and upkeep of new infrastructure and spur economic development. Concrete pavers such as ones currently used in Léogâne are recommended for surfacing main and side street roads and plazas in the city center, as these are pervious to water and easy to repair. For resurfacing peripheral roads, application of lime-based soil stabilizers renders a more reliable pavement with low environmental impact at low cost. Such improvements can serve heavier traffic, including small vehicles moving collected waste to a new 'eco-industrial park' (defined below).

One of the major service points also serves as a multi-modal transportation center: a dedicated hub located on one of Haiti's main highways just a short walk from the civic center. The station will accommodate buses, motorcycles, personal motor vehicles, and bicycles and has a small, attached market. It will be linked by new sidewalks, with pick-up and drop-off areas to improve pedestrian safety. Attached to the waiting area and public restrooms is a biodigester (see below) stationed to also receive large deposits of agro-waste delivered along the main road from outlying areas for transfer to the eco-industrial park.

Waste Management

Waste is an unrealized resource, its management central to local economic revitalization. It creates employment, fosters community involvement and promotes environmental stewardship. Proper refuse management reduces water supply contamination as well. A comprehensive proposal for Léogâne's waste diversion was developed that separates plastics, organics, metal and wood for recycling. It includes a new on-site processing facility, along with storage, transport, and connections to the large scaled biodigester and related generation facility (see An Eco-Industrial Park for Léogâne below). A mechanism to incentivize citizens to deliver sorted materials to the collection and pick-up points will be overseen by a new collection administration system. An educational campaign inculcating the benefits of resource conservation will also be vital for implementation.

Waste-to-Biogas

Seventy-five percent of Haiti's total waste stream is organic, comprised of crop residuals and other biomass, along with food and other domestic waste (Booth et al. 2010). It constitutes a largely untapped resource. The plan calls for a distributed biogas production system made up of strategically placed and variously scaled biodigesters (small for a cluster of dwellings, to larger community-scaled units). These biodigesters rely on slow anaerobic digestion that yields methane from organic human and animal waste mixed with agro-residuals. This is converted into useful biogas and a residual soil amendment byproduct. Benefits include production of a new fuel source for cooking or for additional electricity generation; improved sanitation and public health; greenhouse gas reduction, and increased farm yields, along with job creation.

An Eco-industrial Park for Léogâne

Léogâne sits in an agricultural region long renowned for sugarcane production, Haiti's second largest cash crop after coffee. Yet in recent years, production and profits have faltered and Haiti has become a net importer of sugar from South America (Nienaber 2010). Léogâne now has one of the lowest yields for sugar cane in the Western hemisphere, with a range of 8 to 40 MT per hectare, compared to 85 MT in other parts of Haiti or Latin America (Barjon 2011b). Revival of the sugarcane industry (along with restoration of local food production) is essential for Haiti's overall recovery. Better management of existing resources and development of new markets would restore production to historic levels. A resurgent cane industry will restore much-needed jobs, as two-thirds of the Haitian work force and 25% of GDP production has remained in agriculture (Eberhard et al. 2010).

Léogâne's sugarcane production can be increased through more efficient management of existing resources as well as repair and expansion to the local processing facility, the Darbonne Sugar Mill. Additional resource efficiency can be gained through biodigestion of bagasse, the fibrous by-product of cane processing. The biogas extraction method has a higher energy capacity than simple bagasse combustion, which is a source of pollution. The process also co-produces a high quality fertilizer. Such an integrated resource management approach makes use of unrecognized assets, while reducing reliance on expensive fossil fuel fertilizers and potentially boosting agricultural yields (Piasecki 2013). Extensions to the Darbonne Mill property would effectively turn it into an eco-industrial park. Colocation of an industrial-scaled biodigester at the sugar mill would provide secondary or stand-by biogas cogeneration in Léogâne. The bagasse/biodigester upgrade can be coupled with Léogâne's collected organic waste, delivered and stored on-site. This new co-generation system would have a capacity of 31.1 MW (Brown & Ward 2012). Plastic waste would also be processed as a complementary use in this eco-industrial park, recycled into lightweight compressed bricks for local use.

DISCUSSION

Challenges and Next Steps

This proposal was built around the regulation of distributed renewable energy with the pumped-hydro storage as its lynchpin. To operate effectively, local environmental conditions must be stabilized. In addition construction must be staged to maximize resources. There are technological challenges as well. Pipelines must be protected from corrosion. The PHS water being exchanged requires filtration. The upland reservoirs will also need a measure of protection against sediment accumulation, a serious consequence of deforestation.

Successful ecological restoration will be contingent on temporary or permanent land acquisition. Stable administration and governance are essential for the necessary adjustments to land ownership and/or the creation of easements, complex social and financial transactions for Haiti. For fair distribution of services, placement of infrastructure assets cannot be done piecemeal, but rather systematically and according to a codified system such as GEM. Implementation of GEM's uniformity and modularity will require local, departmental and national cooperation.

A sophisticated understanding of local values and culture will be critical to program delivery. Particularly in Haiti, success or failure will depend upon the quality of local engagement as well as a commitment to local employment. The plan envisions an extensive stakeholder engagement process, which develops infrastructure based on input of local inhabitants, while conveying information about consumer responsibilities and costs. Long-term maintenance of these systems will demand skilled administrative and enforcement personnel. Therefore, a central strategy of GEMi, which will be necessary to implement this proposal, is to help train a workforce for the systems. Creating general awareness of the link between natural resources and infrastructure services is critical. This, along with technical training are GEMi cornerstones, not just for sustainable infrastructure, but ultimately sector-wide to balance short-term needs for water, food and shelter with investment in long-term. Given Haiti's history of electricity "piracy," it is also imperative to maintain the security of infrastructural components. The logistics of installing, maintaining and protecting expensive alternative energy infrastructure demand ongoing staff and security presence.

Creating and maintaining local political will and the capacity to coordinate with national ministries and other resources constitute the primary challenges to the plan's implementation. In-depth meetings with local leaders at all levels of governance and direct engagement with community members will be essential. Education will also be pivotal for acceptance of the paradigm shift involved in cross-sector integration. In many ways, Haitians are ready for this transition, given the failure of many infrastructure investments to date. A related challenge will also be first world buy-in to new infrastructural regimes implied by the integrated and autonomous ecodistrict—one designed with context sensitivity, with local input and for local self-reliance.

Financing will be an obstacle for integrated infrastructure systems, as it requires coordination of resources and personnel from across what are traditionally 'siloed' ministries. Funds sourced from the international community are also needed for the initial outlays, while the program embeds plans for self-sufficiency after initial support is withdrawn. Estimations of lifecycle costs and comparisons between conventional fossil fuel (coal or gas) and the proposed alternative energy systems are not meaningful here since all fuel sources must be imported by this island nation. While currently, PetroCaribe provides a very generous rate to Haiti that allows for a 25 year payback period with 2% interest, (Jacome 2011), the Government of Haiti lacks sufficient credit to ensure a competitive fuel supply for the long-term. Costs for solar PV farms installed in Haiti would likely incur a 30 to 40% premium over U.S. costs due to charges for shipping, tariffs (Gregory2011b) and transport to hard-to-access locations. Installation of a smart-grid energy network is more competitive for Haiti (as other emerging economies that lack existing energy infrastructure) and would provide a significant leap forward for this energy-impoverished nation. Biomass digesters are also likely to be cost effective given the tandem social and economic benefits of also providing a solid waste management system over and above a supplementary energy system. The cost benefits for natural systems regeneration are significant. The outlay for upland ecological restoration is likely to be more than offset by eliminating future loss of life and property from seasonal and hurricane flooding. The natural systems afford a measure of resilience where none currently exists.

CONCLUSIONS

Substandard infrastructure must be redressed in the recovery from major disasters and in the development of sustainable models (Kijewski-Correa et al. 2012). Numerous opportunities exist throughout the island of Haiti and elsewhere for replication of a "whole systems," integrated infrastructural approach. Most of the proposed technologies and strategies are adaptable to areas with available natural resources. In Haiti, where simply meeting the daily needs of the population is an extraordinary challenge, the creation of relatively low cost, integrated, self-reinforcing and resilient solutions can hopefully launch long-term self-reliance and economic development. Elsewhere, Small Island Developing States (SIDS) might utilize similar concepts, adapting them to local geography, meteorology and ecology as a pathway to sustainability. SIDS additionally face enormous stresses from climate instability. Therefore they can be at the forefront in adopting renewable technologies for both climate mitigation and adaptation purposes.

Vulnerability stems from inadequate knowledge as well as deficient resources to apply that information (Kijewski-Correa & Taflanidis 2011). This proposal builds capacity and resiliency through applied practical knowledge of resources, their interdependencies, and by optimizing their reciprocities. Various components of this proposal may find more universal application. Sector integration is both common sense and necessary in areas with limited resources for development. The restoration of ecosystem services, renewable power systems designed around the energy/water/waste nexus, and the use of modular and economic nodal service points, provide a powerful grounding strategy useful in many developing and even developed areas. Though yet untested, it represents a logical, and perhaps revolutionary next step towards sustainability from which the developed world too may draw lessons.

AUTHORS' CONTRIBUTIONS

HB led both the study and design development with the assistance of MW. Both authors read and approved the final manuscript.

ACKNOWLEDGEMENTS

This project was conceived by the corresponding author and its components largely realized through her efforts and those of the teaching assistant working with Master of Science Urban Sustainability students participating in the seminar 'Case Studies in Sustainable Infrastructure' at the City College of New York. The design concept builds upon strategies under development by Daniel C. Gregory of Positive Energies, LLC who co-founded GEM.Responsible editor: Michael Piasecki.

REFERENCES

1. Adelman C (2011) Haiti: Testing the Limits of Government Aid and Philanthropy. Brown J World Aff 17(11):91

2. Archibold R (2010) Haitians Plunge into Muck to Stem Cholera. New York Times.

3. Barjon R (2011) Effectiveness of Aid in Haiti and How Private Investment Can Facilitate the Reconstruction. U.S Senate Subcommittees of Foreign Relations on International Development and Foreign Assistance and Western Hemisphere hearing "Rebuilding Haiti in the Martelly Era". 15

4. Barjon R (2011) Effectiveness of Aid in Haiti and How Private Investment Can Facilitate the Reconstruction. U.S Senate Subcommittees of Foreign Relations on International Development and Foreign Assistance and Western Hemisphere hearing "Rebuilding Haiti in the Martelly Era.". 9

5. Barjon R (2011) Effectiveness of Aid in Haiti and How Private Investment Can Facilitate the Reconstruction. U.S Senate Subcommittees of Foreign Relations on International Development and Foreign Assistance and Western Hemisphere hearing "Rebuilding Haiti in the Martelly Era.". 6

6. Barjon R (2011) Effectiveness of Aid in Haiti and How Private Investment Can Facilitate the Reconstruction. U.S Senate Subcommittees of Foreign Relations on International Development and Foreign Assistance and Western Hemisphere hearing "Rebuilding Haiti in the Martelly Era.". 13

7. Booth S, Funk K, Haase S (2010) Haiti Waste-to-Energy Opportunity Analysis. NREL. 1 http://www.dinepa.gouv.ht/wash_cluster/index.php?option=com_rokdownloads&view=file&task=download&id=1025%3Anrel-haiti-waste-to-energy-final-nov-14-2010&Itemid=27

8. Brown H (2011) Global Energy Model's Pilot Project for Haiti: studies & concepts based on GEM's roadmap for energizing the country. 13 http://globalenergymodel.org/wp-content/uploads/2012/07/Haiti-Presentation-CUNY

9. Brown H, Ward M (2012) Planning an Ecodistrict: Integration of Critical Infrastructure Proposed for the Commune of Léogâne in

Haiti. The City College of New York, CUNY. p 35 http://www. issuu.com/mnward/docs/gemi_leogane_booklet

10. Brown H, Ward M (2014) A Haitian 'ecodistrict:' conceptual 881 design for integrated, basic infrastructure for the commune of Léogâne. 882. Earth Perspectives 1:4

11. Cholera in Haiti. Center for Disease Control and Prevention. http://wwwnc.cdc.gov/travel/notices/watch/haiti-cholera

12. Collier P (2009) Haiti: From Natural Catastrophe to Economic Security. Report for the Secretary-General of the United Nations. 17

13. Dilip Ahuja , Dilip , Tatsutani , Marika (2008) Sustainable Energy for Developing Countries. The Academy of Sciences for the Developing World. Sustainable Energy Rep. 21 http://twas.ictp. it/publications/twas-reports/SustainEnergyReport.pdf

14. Eberhard M, Baldridge S, Marshall J, Mooney W, Rix G (2010) The Mw 7.0 Haiti earthquake of January 12, 2010: USGS/EERI Advance Reconnaissance Team report. US Geological Surv Open-File Rep. 1048 iv

15. Global Energy Model Institute (2012) GEMi | Helping eliminate global energy poverty. http://globalenergymodel.org/

16. Gregory D (2010) The Global Energy Model. A Green Energy Corp White Paper. 10 http://greenenergycorp.com/File/View/ b78fabc6-7247-483b-ac05-1253584d02d7

17. Gregory D (2011a) Global Energy Model. 4-6 http:// globalenergymodel.org/wp-content/uploads/2012/07/GEM-WPv6.pdf

18. Gregory DC (2011b) Global Energy Model (GEM). Positive Energies, White Paper. http://globalenergymodel.org/wp-content/ uploads/2012/07/GEM-WPv6.pdf

19. Haiti - Energy (2012) KOICA handed over two Electric Power Projects to the Government. Haiti Libre. http://www.haitilibre. com/en/news-6206-haiti-energy-koica-handed-over-two-electric-power-projects-to-the-government.html

20. Holm D (2005) Renewable Energy Future for the Developing World. In: International Solar Energy Society. Germany: ISES. p 2

21. Holm D, McIntosh J (2008) Renewable energy- the future for the developing world. Renew Energy Focus 9:1

22.   Jacome F (2011) The Current Phase of Venezuela's Oil Dipolmacy in the Caribbean. Friedrich Ebert Stiftung. http://library.fes.de/pdf-files/bueros/la-seguridad/08723.pdf

23.   Kijewski-Correa T, Taflanidis A (2011) The Haitian housing dilemma: can sustainability and hazard-resilience be achieved? Bull Earthquake Eng. 7 DOI: 10.1007/s10518-011-9330-y

24.   Kijewski-Correa T, Taflanidis A, Mix D (2012) Empowerment Model for Sustainable Residential Reconstruction in Léogâne, Haiti, after the January 2010 Earthquake. Leadersh Manag Eng 12:286

25.   Klarreich K, Polman L (2012) The NGO Republic of Haiti. Nation.

26.   Koehler R, Mann P (2011) Field Observations from the January 12, 2010 Haiti Earthquake: Implications for Seismic Hazards and Future Post-Earthquake Reconnaissance Investigations in Alaska.   12   http://www.dggs.alaska.gov/webpubs/dggs/ri/text/ri2011_002.PDF

27.   Lemons D (2012) At this writing, GEMi has been named by the Haitian Minister Delegate for Energy Security as the co-developer of a master plan for energizing the nation, fashioning a national grid based upon upgrades and additions to existing resources, implementation of new generation technologies, the roll-out of rural 'eco-districts', and the developer of a ten-year plan for ramping up education and training for the energy sector. Personal communication.

28.   Library of Congress –Federal Research Division (2006) Country Profile: Haiti. 10 http://lcweb2.loc.gov/frd/cs/profiles/Haiti.pdf

29.   Metzger J, Olsson A (2013) Sustainable Stockholm: Exploring Urban Sustainability in Europe's Greenest City. 200

30.   Nienaber G (2010) Bill Clinton Puts Influential Muscle Behind Agricultural Production in Haiti. Huffington Post. http://www.huffingtonpost.com/georgianne-nienaber/bill-clinton-puts-influen_b_697906.html

31.   O'Connor R (2012) Two Years Later, Haitian Earthquake Death Toll in Dispute. Columbia J Rev. http://www.cjr.org/behind_the_news/one_year_later_haitian_earthqu.php?page=all

32.   Piasecki M (2013) Personal communication.

33. OECD (2006a) Promoting Pro-poor Growth Infrastructure. http://www.oecd.org/development/povertyreduction/36301078.pdf

34. OECD (2006b) Promoting Pro-poor Growth Infrastructure. 13 http://www.oecd.org/development/povertyreduction/36301078.pdf

35. Rebuilding Léogâne Planning Workshop (2011a) Master Plan. "Partie 2: Plan de reconstruction post-séisme de la commune de Léogâne.". 2-8 http://www3.nd.edu/~ce4haiti/Information%20Database/MasterPlan.pdf

36. Rebuilding Léogâne Planning Workshop (2011b) Master Plan. "Partie 2: Plan de reconstruction post-séisme de la commune de Léogâne.". 2-2 http://www3.nd.edu/~ce4haiti/Information%20Database/MasterPlan.pdf

37. Shepard W (2010) Streets of Haiti are Dangerous. http://www.vagabondjourney.com/streets-of-haiti-are-dangerous/

38. Sontag D (2012) Rebuilding in Haiti Lags After Billions in Post-Quake. New York Times.

39. The Reef Ball Foundation-Designed Artificial Reefs. http://www.reefball.org/

40. University of Notre Dame (2011) Rebuild Léogâne: Community Planning Workshop Report. 45 http://haiti.nd.edu/assets/54558/

41. University of Notre Dame (2011) Rebuild Léogâne: Community Planning Workshop Report. 27 http://haiti.nd.edu/assets/54558/

42. University of Notre Dame (2011c) Rebuild Léogâne: Community Planning Workshop Report. 18 http://haiti.nd.edu/assets/54558/

Citations

CHAPTER 1

Chinedu I. Ossai, "Advances in Asset Management Techniques: An Overview of Corrosion Mechanisms and Mitigation Strategies for Oil and Gas Pipelines," ISRN Corrosion, vol. 2012, Article ID 570143, 10 pages, 2012. doi:10.5402/2012/570143.

CHAPTER 2

Wren Montgomery, Mark A. Sephton, Jonathan S. Watson, Huang Zeng, and Andrew C. Rees, Minimising Hydrogen Sulphide Generation During Steam Assisted Production of Heavy Oil, doi:10.1038/srep08159.

CHAPTER 3

Emma E H Doyle, Douglas Paton, and David M Johnston, Enhancing Scientific Response in a Crisis: Evidence-Based Approaches from Emergency Management in New Zealand, doi:10.1186/s13617-014-0020-8.

CHAPTER 4

Islam Safak Bayram and Ioannis Papapanagiotou, A Survey on Communication Technologies and Requirements for Internet of Electric Vehicles, doi:10.1186/1687-1499-2014-223.

CHAPTER 5

Bhzad Sidawi and Abdulsalam Alsudairi, The Potentials of and Barriers to the Utilization of Advanced Computer Systems in Remote Construction Projects: Case of the Kingdom of Saudi Arabia, doi: 10.1186/2213-7459-2-3.

CHAPTER 6

Boukelia Taqiy Eddine and Mecibah Med Salah, Solid Waste as Renewable Source of Energy: Current and Future Possibility in Algeria, doi: 10.1186/2251-6832-3-17.

CHAPTER 7

Hillary Brown and Miriam N Ward, A Haitian 'Ecodistrict:' Conceptual Design for Integrated, Basic Infrastructure for the Commune of Léogâne, doi: 10.1186/2194-6434-1-4.

Index